Sustainable Engineering Forum 2017

AA008971

Topical Conference at the 2017 AIChE Spring Meeting and 13th Global Congress on Process Safety

San Antonio, Texas, USA
26 – 30 March 2017

ISBN: 978-1-5108-4140-6

Printed from e-media with permission by:

Curran Associates, Inc.
57 Morehouse Lane
Red Hook, NY 12571

Some format issues inherent in the e-media version may also appear in this print version.

Copyright© (2017) by AIChE
All rights reserved.

Printed by Curran Associates, Inc. (2017)

For permission requests, please contact AIChE
at the address below.

AIChE
120 Wall Street, FL 23
New York, NY 10005-4020

Phone: (800) 242-4363
Fax: (203) 775-5177

www.aiche.org

Additional copies of this publication are available from:

Curran Associates, Inc.
57 Morehouse Lane
Red Hook, NY 12571 USA
Phone: 845-758-0400
Fax: 845-758-2633
Email: curran@proceedings.com
Web: www.proceedings.com

TABLE OF CONTENTS

482030 AGILE Award Keynote Address

Abstract:

Factors in Selecting AGILE Honoree

1. Overall sustained excellent industry leadership in businesses that employ chemical engineers

2. Recognized for efforts in innovation and creating strategies that ensure success

3. Value of the keynote to the meeting attendees and the members of the Institute Schedule

Event	Time	Location
Continental Breakfast	7:00AM - 8:00AM	Henry Gonzalez Convention Center, Lila Cockrell Theatre
AGILE Keynote Address	8:00AM - 9:15AM	Henry Gonzalez Convention Center, Lila Cockrell Theatre

The 2017 AIChE Spring Meeting and 13th GCPS AGILE Keynote Address will be presented by AGILE award recipient John Y. Televantos, Partner, Arsenal Capital Partners.

John Televantos is a Partner at Arsenal Capital Partners, a middle market private equity firm focused on the specialty industrial, healthcare and specialty financial service sectors. Before joining Arsenal in 2006, Dr. Televantos was President of Aqualon Company, owned by Hercules, Inc.; President and Founder of Helios Chemical Company; and Chief Executive Officer for Foamex International.

Dr. Televantos began his career with Union Carbide in 1977 in silicones research. There he developed two key processes for the manufacture of Methyl Chloride and Siloxane polymers that were commercialized. These processes not only improved efficiency and reduced cost but also dramatically reduced waste. He moved through a variety of management positions there and later at ARCO Chemical, where he became vice president of strategy and development.

At ARCO John co-invented and led a team that developed and commercialized a breakthrough technology in polyol manufacture (a key raw material for polyurethanes) eliminating a whole process step along with waste and emissions, while at the same time opening the door to the creation of new higher-performance polyurethanes. The technology has since been licensed by most of the major global polyol manufacturers with significant waste reduction worldwide.

At Arsenal Capital, Dr. Televantos serves as co-head of the specialty industrial practice. During his tenure, he has been part of several acquisitions and successful exits in the chemicals and materials sectors. In addition to serving on the boards of several Arsenal-owned companies, he is lead director on the board of Balchem Corp.

Copyright © American Institute of Chemical Engineers. All rights reserved.

480703 Big Data Analytics Skills - a Revolution Lacking in Revolutionaries

Abstract:

Since the advent of the common use of the term "Big Data[i]" in 2010 to describe large, *unruly* sets of data, and the artful analysis thereof, industry has felt besieged by opportunity. The chemical and refining industry has been faced with "Big Data" long before it was cachet. Yet within industry there is still a struggle to get hearts and minds around what the Big Data Journey means to an organization and how to take advantage of the data opportunities that lie within. The added complication to the chemical industry is that it must add "contextualization" to the data analysis because it's not enough just to know there is a signal, we need to know what the signal means[ii]. Yet during this data revolution there is an underlying challenge that Academia, Industry and Governmental Agencies must tackle together to ensure the US Chemical industry retains its competitive position in the world. Together we need to provide education for the higher level skills and abilities required across roles in manufacturing to assure that no one gets left behind in the new Big Data age. Indeed, the next great technological "step" made may not be in engineering design or IoT connectedness (reference Industry 4.0), it may be in designing and creating a better human capability in "data acuity" out of our university system; operators, engineers and scientists alike who are better prepared for the data age than their predecessors. This talk will take a look at what Industry requires in skills sets and perspective to be "data ready", and what Academia and Government might do to improve the output of data-ready skills.

---

[i] Big Data, What is it? White, Chemical Engineering Progress, AIChE March 2016.

[ii] Big Data, Getting Started on the Journey. Colegrove, Seasholtz, Khare, Chemical Engineering Progress, AIChE March 2016.

Copyright © American Institute of Chemical Engineers. All rights reserved.

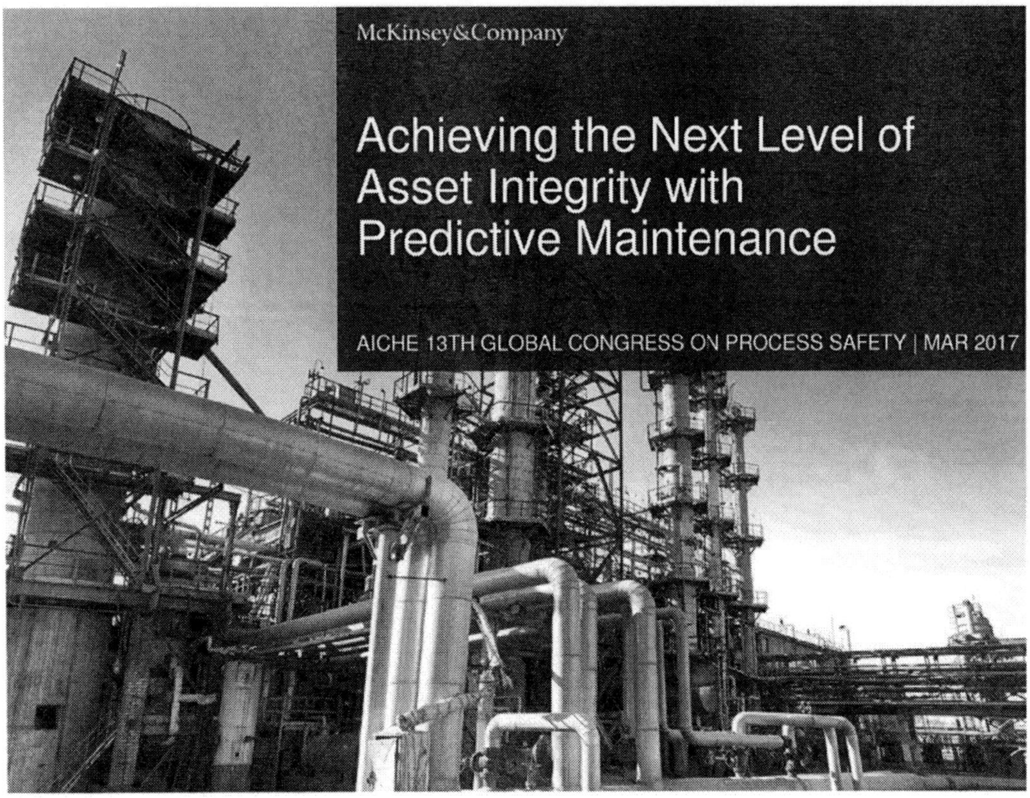

Abstract

Achieving the next level of asset integrity with predictive maintenance

Exceptional organizations manage to deliver superior operational performance without compromising on safety and integrity. On the contrary, high performing players are at the forefront of equipment reliability for critical equipment and safety performance.

In this context, introducing Predictive Maintenance can dramatically help achieve the next level of asset integrity by leveraging data - structured and unstructured, machine and non-machine based – in order to generate insights with advanced analytics methods that would not be possible with conventional techniques
Implementing Predictive Maintenance generates three benefits that reinforce one another:
Predict when equipment will fail and help reduce the safety and integrity threats
Avoid failures and positively contribute to reduce the scale of the value at risk
Enhance specific asset knowledge and learn more about the behavior of critical equipment

In a recent client project for an upstream oil and gas facility, the main gas compressors have been identified as **critical equipment**: they account for many failures and a large portion of production losses. A Weibull analysis showed that out of several known failure modes, some show a random failure pattern, leading to sudden breakdowns releasing high energy and thus representing a safety issue. As per the random failure behavior, classical time- or usage-based maintenance strategies are not fruitful.

Combining 100's of gigabytes of loss data, maintenance work order data and sensor data from 1,200 sensors around the equipment, a data driven model has been created showing that out of the 1,200 sensor sources, 43 explanatory variables predict the time to failure with ~80% accuracy. The model has been tweaked further in order to reduce the number of false negatives while accepting by this a certain number of false positives (i.e. false alarms) due to safety reasons.

As a result of this model, an early warning of an imminent failure means repairs can be planned, spares can be gathered and the equipment can now be shut down in a safe and orderly manner. Overall, this represents an average reduction from 14 to 4 days of downtime per breakdown event.

Outline

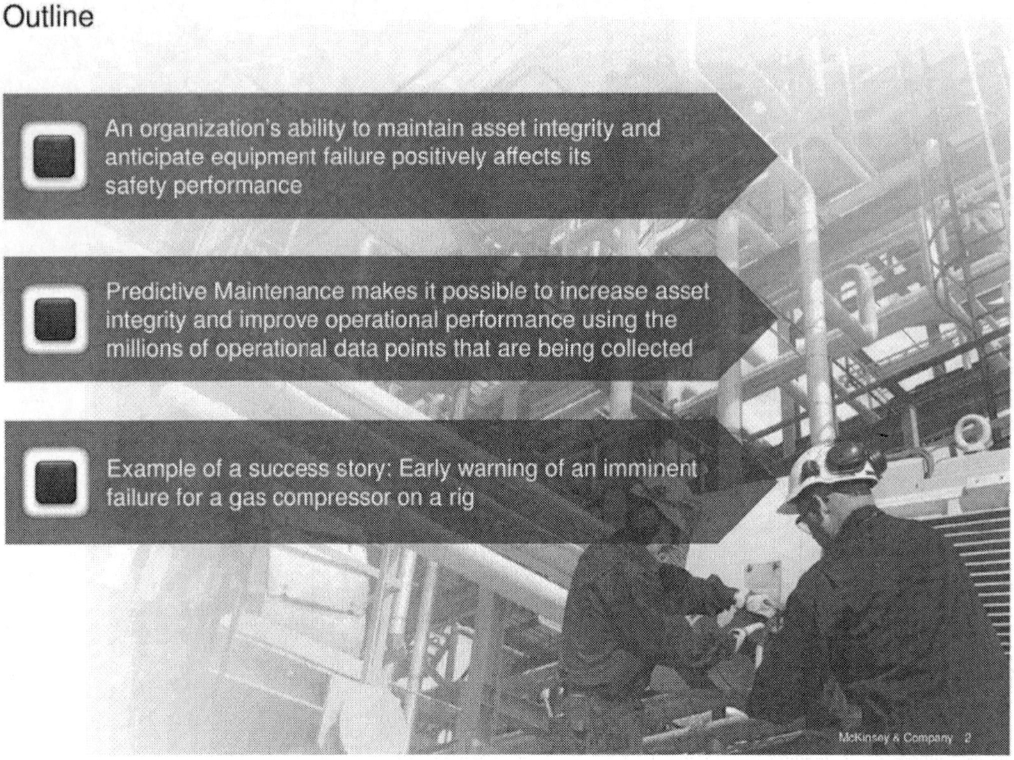

- An organization's ability to maintain asset integrity and anticipate equipment failure positively affects its safety performance
- Predictive Maintenance makes it possible to increase asset integrity and improve operational performance using the millions of operational data points that are being collected
- Example of a success story: Early warning of an imminent failure for a gas compressor on a rig

McKinsey & Company 2

There is a strong correlation between equipment reliability and safety performance

1993 - 2003 (base 100)

EXAMPLE FROM ALUMINUM SMELTING

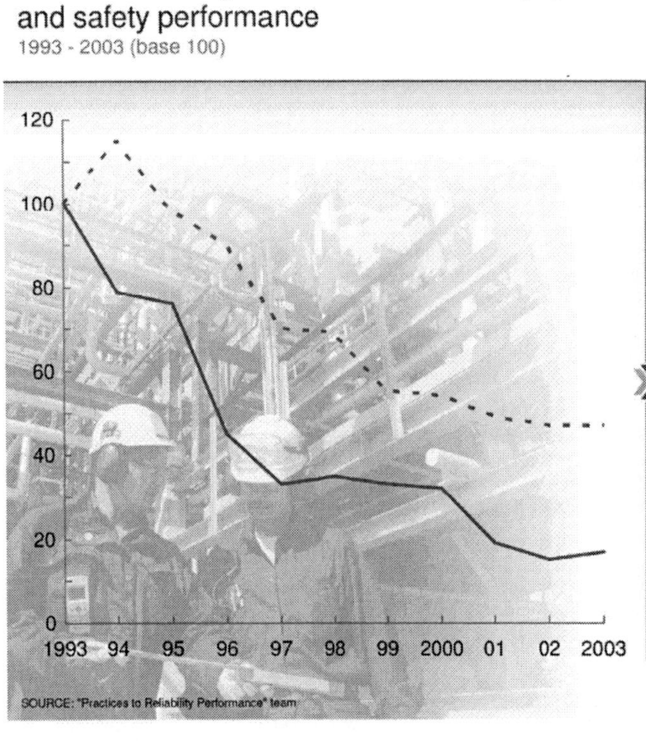

- **Safety** and reliability performance require **control over the plant environment**
- **Unscheduled work** – often in rush – **is by nature much more risky** than planned interventions
- **Reliability and safety performance share multiple common drivers**
  - Importance of **behavioral factors**
  - **Anticipation rather than reaction as a management practice**
  - **Ownership** and care for equipment

SOURCE: "Practices to Reliability Performance" team

McKinsey & Company 4

Industry collects Petabytes of data which can be used for Predictive Maintenance to improve asset integrity and reliability

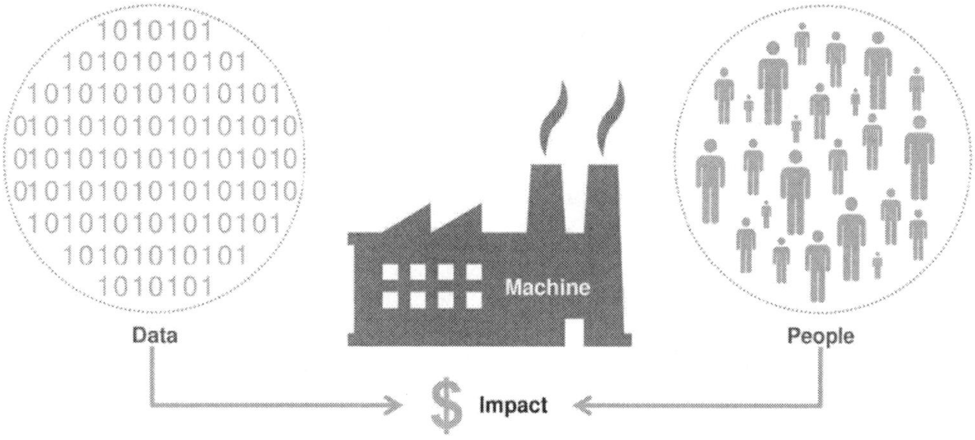

SOURCE: McKinsey

Getting from data to impact requires a combination of many different skills

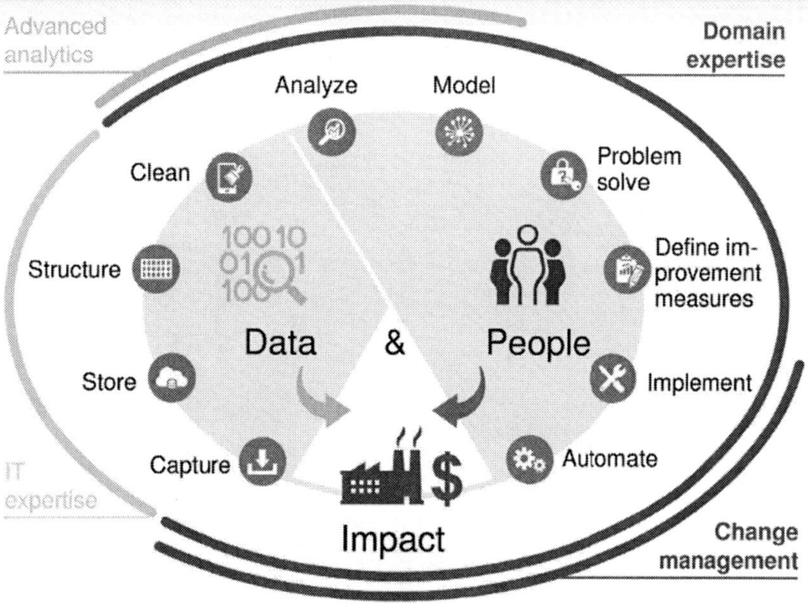

SOURCE: McKinsey

Advanced Analytics generates previously untapped insights in Predictive Maintenance from troves of historical data

Predictive Maintenance
is the use of data (structured and unstructured, machine and non-machine based) to generate insights with advanced analytical methods that would not be possible with conventional techniques

Predict

Failure forecasting: Model future asset behavior to anticipate needs *(e.g., HR, spare parts inventory, capital replacement)*

Incipient failure detection: Continuous data acquisition & analysis to detect or predict failure *(e.g., early warnings to allow planned intervention)*

Avoid

Avoid failures

Identification of undesired patterns in data as primary causes of failure (e.g., preventive actions in specific locations of roads with poor conditions to prevent truck failures)

Learn

Leverage specific asset knowledge

Access reliability data from a global equipment database comprised of hundreds of thousands of components across the globe *(e.g. Failure rates, failure mode distributions, causes)* to inform maintenance strategies

SOURCE: McKinsey

McKinsey & Company 7

The main benefits of implementing predictive maintenance include availability increase and cost reduction

Predictive maintenance aims to ...

... detect early signs of problem **for timely correction at minimal costs**

... prioritize and optimize maintenance resources using real-time data

... increase unit availability due to **shorter downtimes**

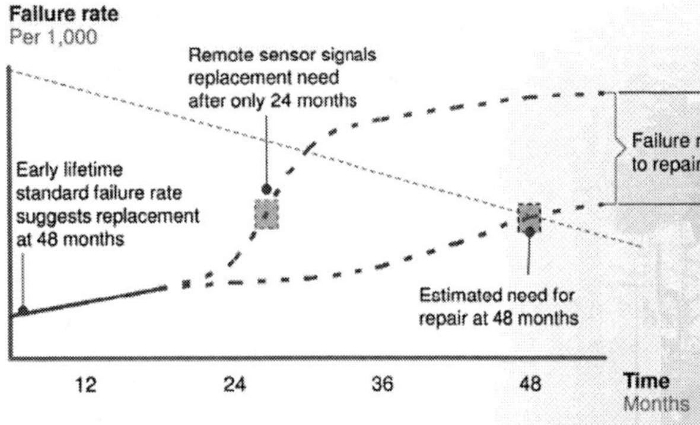

Typical impact of PdM

Machinery breakdowns	-50%
Spare part inventory	-20%
Total machine downtime	-50%
Overtime expenses	-20%
Machine life	+20 - 40%
Overall productivity	**+20 - 30%**

SOURCE: ARC Advisory Group; Department of Defense; expert interviews; GE; Schaeffler; VibroSystM; WiHART Systems

McKinsey & Company 8

Predictive maintenance is proven to add value in many industries

What was Predictive Maintenance used for

✓ Focus of the engagement
Detailed in following pages

Industry	*Reduce operating costs*	*Reduce maintenance costs*	*Increase machine availability*	Impact
1 Mining			✓	▪ 25% downtime reduction
2 Mining *(Copper mine)*	✓	✓		▪ Savings of ~12% in operational expenditures
3 Oil and gas *(Rig operator)*		✓		▪ Reduced maintenance spend by 18 to 27%
4 Oil and gas			✓	▪ ~70% reduction in downtime on critical component
5 Railroads			✓	▪ Reduced failure cost by 20% to 30%
6 Heavy-Motors *(Cruise operator)*	✓	✓		▪ Reduced operating cost by 7%
7 Semiconductor	✓			▪ Increased failure prediction to 87%

We have more than 20 use cases. The remaining cases cannot be shared due to confidentiality.

SOURCE: McKinsey Asset Productivity Service Line

Summary of the case

Applying advanced analytics in off-shore oil and gas gives early failure warning can reduce production losses due to gas compressor by ~70%

Situation

Sector and site
- Offshore oil and gas operator
- Rig maintenance

Situation
- Gas compressors are **critical equipment**: they account for many failures and a large portion of production losses
- Loss data, maintenance work order data and sensor data is collected

Question
- To what extent can we reduce maintenance spend and improve reliability by predicting when a failure will happen?

Typical failure mode: inlet strainer

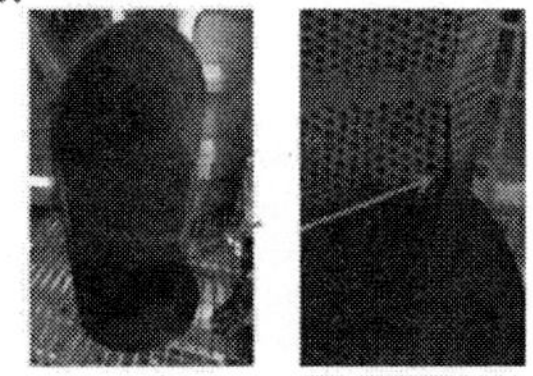

Key Activities

Data collection from several sources
- 120 gigabytes of data
  - ~1,200 sensors
  - 10 years of work orders and deferment data

Identify failures and construct target variable
- Identify if downtime is caused by compressor

Compr | Other | t

Create a model to predict time to failure
- For any point in time we predict the **time to the next failure**

Inputs 1,200 sensors → Prediction Model → **Time to next failure**

- **Inputs** read from 1,200 sensors
- **Regression Model** predicts the time to failure with ~80% accuracy

Impact

- Downtime due to gas compressor is expected to reduce by **~70%** by planning intervention when a failure is imminent
  - **Prediction of time to failure:** if intervened at predicted TTF < 5 days, you would act between 9 and 6 days before the failure actually will happen

Predicted TTF for different failures, Days

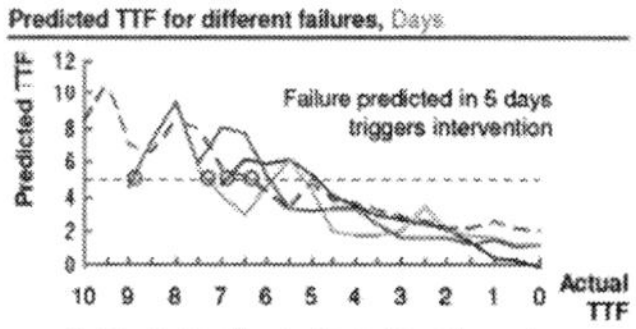

  - **Early warning of an imminent failure** means repairs can be planned. Reduction is on average from 14 to 4 days of downtime

		Breakdown		
Old situation	Up	Down 14 days	Up	
New situation	Up	Prep.	Down 4 days	Up

1 TTF = Time To Failure, each of the lines in the graph is one failure case

SOURCE: McKinsey

Details of the case

Data collection was based on 120 GB of data – including more than 1200 sensors and 10 years of work orders

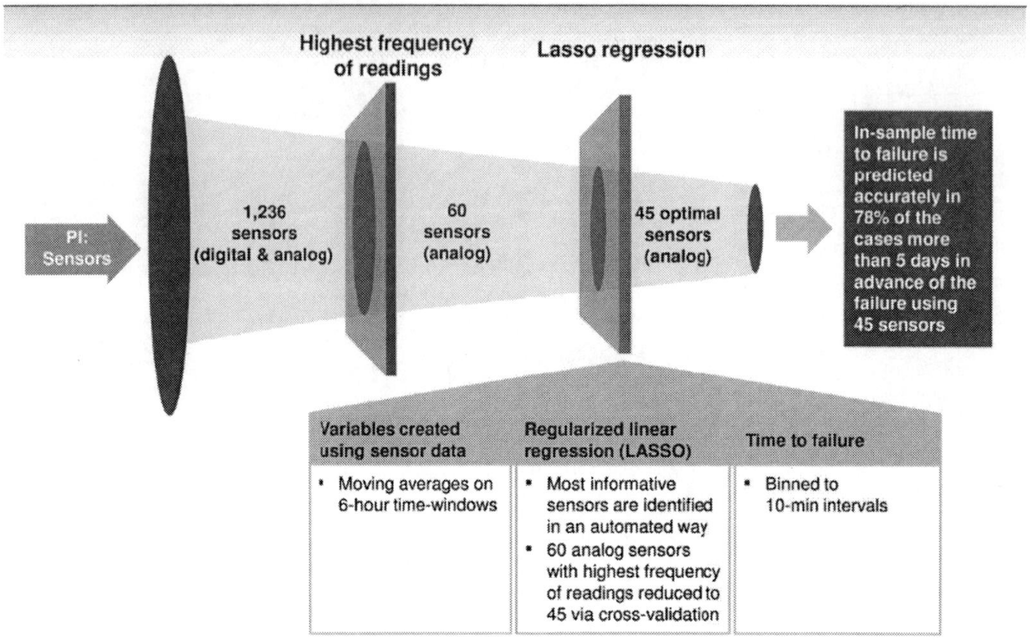

SOURCE: McKinsey

Details of the case

In order to predict downtimes, data preparation as well as a clear target variable were keys to success

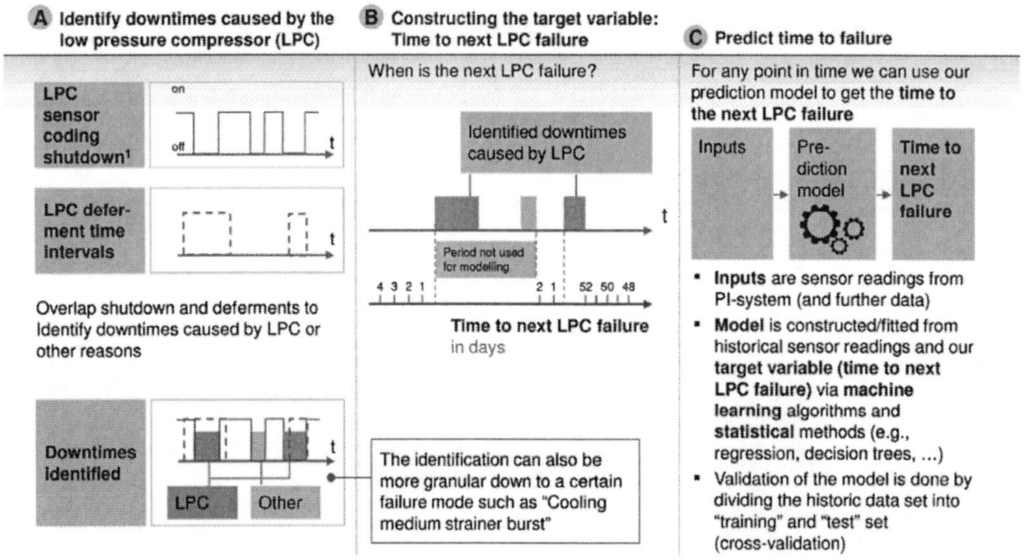

SOURCE: McKinsey

FROM ANALYSIS TO VALUE CREATION

Top down value assessment for predictive maintenance

PRELIMINARY

Digital asset management

- ↑ **Availability 2-5%**
  - ↓ **Planned downtime 10-20%**
    - Decrease in work order frequency
      - Work order cadence to equipment needs instead of OEM recommendation
    - Decrease in work order duration
      - Word order instructions pushed to mobile devices
      - Improved preparation to increase employee wrench time
      - Fully digital and mobile enable maintenance process and cycle
  - ↓ **Unplanned 30-50%**
    - Actively performing routine work orders based on predictive analytics
    - Monitoring conditions of equipment to accurately predict failures
- ↓ **Maintenance cost, 5-25%**
  - ↓ **Operating maintenance costs**
    - Cheaper, routine work orders (predictive > reactive)
    - Improve part and tool inventory strategy with improved database
  - ↓ **Sustaining capex costs**
    - Longer equipment life through planning component replacement extension
    - Adjust to a smaller fleet size

McKinsey & Company 14

I The predictive model combines sensors in a way that cannot be replicated manually

Most important sensors for prediction

10 most significant out of a total of 45 sensors

Rank	Sensor	Unit of measure	Relative contribution Abs. value of coefficient
1	SEAL VENT	BARG	4.2
2	SEAL GAS VENT DP	BAR	2.3
3	FUEL GAS PRESSURE	BARG	1.9
4	COMPRESSOR FLOW DP	MBAR	1.8
5	INLET FILTER DP	MBAR	1.6
6	SEAL GAS SUP PRESS	BARG	1.3
7	ENG BRG2/3 DRAIN	DEG C	1.3
8	FUEL SYS AIR SUPPLY	BARG	1.2
9	FUEL GAS FLOW DP	MBAR	1.1
10	PRI NDE SEAL VENT	BARG	1.1

- The statistical model has identified a total of **45 sensors** to be significant when predicting compressor failures
- The contribution of the sensors is evenly distributed – there is **no one sensor that can replace the predictive model**, which combines a total of 45 sensors
- The absolute value of a coefficient can only be extracted in a linear regression case, and that other metrics should be used for other models e.g. variable importance in random forests

SOURCE: Team analysis, Sensor data

McKinsey & Company 15

II We are able to predict breakdowns at least 5 days before failure

Predicted time until failure. 12-hour average of prediction

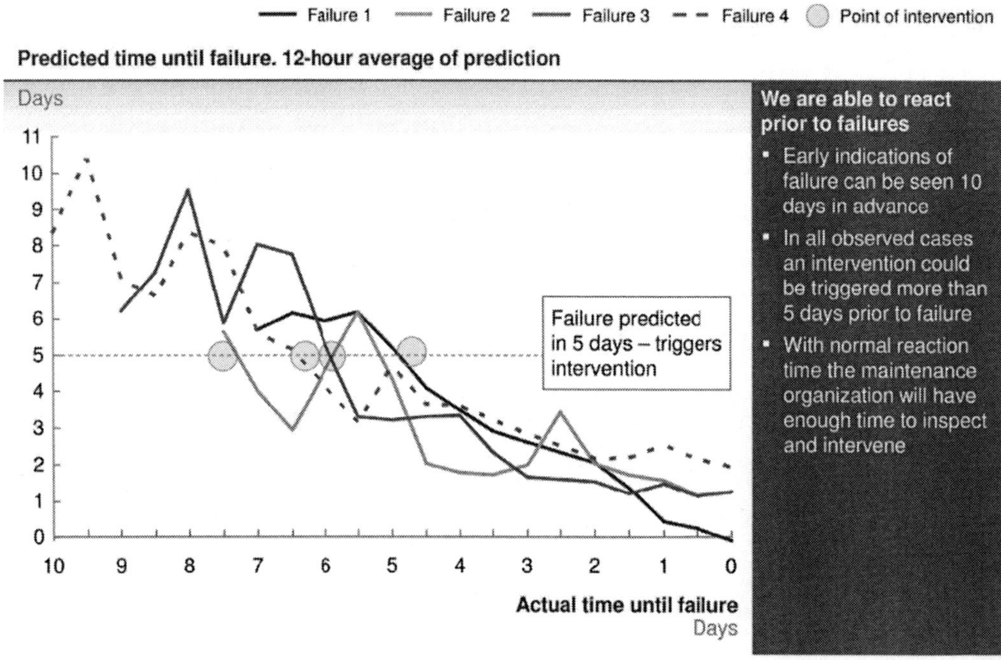

We are able to react prior to failures

- Early indications of failure can be seen 10 days in advance
- In all observed cases an intervention could be triggered more than 5 days prior to failure
- With normal reaction time the maintenance organization will have enough time to inspect and intervene

SOURCE: Team analysis, client data

McKinsey & Company 16

482653 Why Sustainability Is Important at Conocophillips

Abstract:

In today's low oil price environment, it might be surprising to learn that the nation's largest independent oil producer is still actively pursuing sustainability. Sustainable development is about conducting our business to promote economic growth, a healthy environment and vibrant communities, now and into the future. Sustainable development at ConocoPhillips encompasses our environmental performance, stakeholder relations, and responsible operations. To achieve true sustainability, benefits should extend to economic, environmental and social performance. The impact of sustainability will be discussed through an overview of case studies ranging from improved water management to reducing emissions to show how sustainability is smart business. The presentation will show how sustainability is an integral part of ConocoPhillips and covers the life of an asset, from new country entry to decommissioning.

Copyright © American Institute of Chemical Engineers. All rights reserved.

482820 How Product Stewardship Can Enable Product Sustainability, Life Cycle Thinking and Risk Reduction for Business

Abstract:

Sustainability is an essential element of business from the products we make to the way we make them. As we develop or improve products that makes the world a better place, it is clearly important to do so in a responsible manner. All new and significantly modified existing products sold by Owens Corning undergo a comprehensive Product Stewardship review to ensure that they perform as claimed and are both safe and environmentally sound to make, use and dispose. Product reviews address all elements of Owens Corning's Environmental, Health, Safety and Product Stewardship Policy, and are carried out by highly qualified Product Stewardship Review Board members. Reviews are conducted at various stages, including the design, development, test market, manufacture and distribution phases. A Sustainability Mapping Tool is used to evaluate how the new product or process will impact the Company's sustainability goals and to drive decisions in the design phase that will achieve a portfolio of more sustainable products. In this session we will give an overview of how Product Stewardship enables product sustainability, life cycle thinking and risk reduction for business.

Copyright © American Institute of Chemical Engineers. All rights reserved.

482651 Sustainability Drivers and the New Trump Administration

Abstract:

This presentation will begin by summarizing drivers that motivate industry to adopt more sustainable practices and principles ahead of mandatory command and control regulation. Beginning with basic risk management principles, the presentation will review developments that formed the basis for sustainability concepts including evolution of property and other rights of action, regulation of waste, protection of resources, increased scientific knowledge regarding impacts to the environment and public health, increased population density, greater access to information and resulting transparency. The presentation will continue with a summary of current sustainability principles, voluntary programs and developments, such as guidance from the Securities and Exchange Commission regarding material risk and increased shareholder activism, serving to motivate corporations to move forward with revised business models incorporating sustainable decision making, as well as critical developments arising with and after the 2015 Paris Climate Conference which hosted the 21st Conference of the Parties (COP21) and the Sustainable Innovation Forum 2015. Finally, this presentation will review how sustainability programs, incentives and other drivers toward implementation may be impacted by the new Trump Administration policies and early direction.

Copyright © American Institute of Chemical Engineers. All rights reserved.

479971 Big Data Vs. Little's Law: A Systems-Based Framework Changing the Risk-Reward of Large Scale Facilities

Abstract:

The current rush to adopt Big Data ignores the fact that *the data is not the system*. Ignoring system dynamics in the interpretation of thedata often produces the opposite of the intended result. Fortunately, Little's Law ($L = \lambda W$) provides a framework to properly interpret the dynamic nature of data and diagnose the true underlying systemic problems. Little's Law is especially useful in identifying and eliminating *hidden factories* in the system that erase profits, increase systemic risk, and erode morale and culture. Dr. Carrier will explore examples from high-risk environments that have saved hundreds of millions of dollars per year while improving safety. In addition, the amount of data required to make and implement informed decisions is dramatically decreased through the use of Little's Law. The approach will be demonstrated with several examples in Oil and Gas, Refining, and Fine Chemicals worked on by the author. The methodology will also be demonstrated for use in maintenance management and the radical improvement of a mid-sized refinery's MoC (management of change process). Finally, the role of culture in actively the dynamic response of our systems will be discussed.

Copyright © American Institute of Chemical Engineers. All rights reserved.

479057 Big Data Doesn't Necessarily Mean Big Intelligence

Abstract:

Big data is the IT phrase of the moment in almost every industry. The concept of collecting huge amounts of data to learn, manage and drive your business forwards is attractive to all managers, especially as without it you may feel that you're being left behind. But without knowing how to collect data effectively, knowing what data has value, and how to effectively and efficiently use the volume of information, Big Data can be a hindrance, not a help. In Process Safety, we rely on data coming from the field in many instances. Collecting this data can be time consuming and costly, so making the most of the data you have is paramount, as is ensuring the data you collect is seen as a valuable asset by everyone in your organisation. But turning that data into intelligence to guide you on your path to Process Excellence is what it's really all about. Data is as valuable an asset as any other. Look after it. Nutrue it. Value it. And make sure you use it. In this paper we examine the core principles of making PSM data work for you to give you the best usable, actionable intelligence possible.

Copyright © American Institute of Chemical Engineers. All rights reserved.

Meeting the Future Organizational and Technical Training Challenges of a Changing Workforce

Dr. Stewart W. Behie
Manager Safety and Process Risk
Occidental Oil and Gas Corporation
Suite 110, 5 Greenway Plaza, Houston, Texas 77046
stewart_behie@oxy.com

Mr. Matthew K. Henwood
Manager Technical Training
Occidental Oil and Gas Corporation
Suite 110, 5 Greenway Plaza, Houston, Texas 77046
matthew_henwood@oxy.com

Prepared for Presentation at
American Institute of Chemical Engineers
2017 Spring Meeting and 13th Global Congress on Process Safety
San Antonio, TX
March 26 – 29, 2017

Disclaimer: The opinions and belies expressed herein are strictly those of the authors and do not necessarily reflect those of Occidental Oil and Gas.

Meeting the Future Organizational and Technical Training Challenges of a Changing Workforce

Dr. Stewart W. Behie
Manager Safety and Process Risk
Occidental Oil and Gas Corporation
Suite 110, 5 Greenway Plaza, Houston, Texas 77046
stewart_behie@oxy.com

Mr. Matthew K. Henwood
Manager Technical Training
Occidental Oil and Gas Corporation
matthew_henwood@oxy.com

Keywords: technical talent, training and competency assessment, staff retention; multi-generational workforce;

Abstract

The workforce of the future will be dramatically different from that of today in several significant ways. In order to be successful in attracting and retaining technical staff in the future, Companies will have to gain an in-depth understanding of the nature of this technical workforce and be prepared to meet their needs and to match the demands of industry starting now. During most of the 20th century there were three generations actively participating in the workplace during which period older workers were generally the bosses and younger workers toiled under their guidance in a relatively structured manner. The workplace of today has gained a complexity never seen before, as for the first time in history in North America, workers from four distinctive and different generations will be working together. This dynamic will present new challenges and opportunities that need to be managed to achieve success. For example, by 2025, it is reported [(2)] that seventy-five percent of the workforce will be Millennials with their unique perspectives on the workplace, workplace interactions and their place in society. Millennials or Gen Y workers will be working alongside Baby Boomers, as well as Gen X, and Gen Z (Centennials). The workplace is changing fast and successful companies, which will rely on younger professionals to lead them into this uncertain future, must develop systems and structures that considers trans-generational needs of their workforce. All of these factors must be appreciated if companies are to attract and retain talented staff. Training and development programs will become a differentiator among companies in the future if they are to be successful. Attrition is an additional, significant challenge that can lead to accelerated departure of expertise at a time when the oil and gas industry is already experiencing a declining workforce resulting in part from the difference in focus among the different generations. Learning solutions must provide accelerated development compared to "traditional" methods (used for Boomers and early Gen X) and these must be repeatable, measurable and sustainable. The training environment and overall proposition has changed. Blended approaches to learning is the only solution and must reflect the reality that adult learners learn by doing.

This changing workplace profile will be taking place within a looming manpower shortage and potential crisis in all technical skills areas [1]. For example, high and very high skills gaps are predicted to occur in many developed countries including those in North America, and in the fields of engineering and construction starting as early as 2020 [2]. To be successful in this changing world, companies must meet the additional challenges of developing and promoting new approaches to attract and retain the dwindling technical resource of the future. These approaches could include a range of considerations such as: promoting virtual internships and work schemes; developing innovative ways to integrate Human Resources (HR) considerations into corporate strategy development; creating growth opportunities to retain staff; extending employment opportunities for women and older employees on their terms; and fostering talent mobility. All of these approaches, in and of themselves, will present substantial challenges particularly in face of the recent trend towards nationalization and away from globalization.

This paper will examine all of these factors and discuss ways to use these insights to optimize learning processes and customize training programs to match the needs and requirements of the future workforce. This paper will suggest novel approaches to training, skill set development and competency assessment. Finally, this paper will profile one company's approach to understanding the changing future workforce and reflecting this knowledge in technical training, skills set development and competency assessments in order to maintain a position as a leader in the oil and gas sector through innovation, integrity and prudent investment. The challenges of the future are many and will only be met by those who understand them and put in place plans to address them.

1 Introduction

History has shown that to be successful a company in the energy sector they must be able to attract and retain technical staff on an ongoing basis. The energy sector has faced and will continue to face a period of difficulty in this regard. The demographic gap has effectively produced an associated gap in the talent pipeline around the mid-career sector. A company in the energy sector must benchmark senior talent capability and provide rapid solutions to close the early to mid-career gap while also addressing advanced technical skillsets in order to make the solutions sustainable. In a fast changing world this is a real challenge for a number of reasons including the changing profile of generations over time in terms of their personal focus, interests and overall requirements and demands of the workplace.

During most of the 20th century there were three generations actively participating in the workplace; older workers were generally the bosses and younger workers toiled under their guidance in a relatively structured manner. The workplace of today has gained a complexity never seen before. For the first time in history in North America, workers from four different generations will be working together. This dynamic will present new challenges and opportunities that need to be managed to achieve success. By 2025 [2], it is reported that seventy-five percent of the workforce will be Millennials with their unique perspectives on the workplace, workplace interactions and their place in society. Companies that are successful at attracting and retaining the best and brightest talent have and will continue to have a distinct economic advantage over the competition. However, this advantage will become more difficult to achieve and maintain over time. In order to maintain this advantage. It is critically important for companies not only to understand the dynamics of the changing workforce but also to establish organizational structures and programs that will be able to adjust to meet the changes in the technical workforce of the future. This process should start now since significant change has already started. Statistics from the US Census Bureau[18], indicate that the oil and gas industry will experience the departure of about 22,000 petro-technical professionals over the next five years in North America alone while attracting about only 17,000 professionals. Even allowing for the impact of technological efficiencies of increased productivity and operational capability, the industry will fall 20% short of maintaining status quo compared to current conditions. The solution is not only to drive accelerated development but also to provide a viable career proposition to attract new talent and create a diverse and growing workforce.

The situation will soon be further complicated and severely constrained by the looming crisis globally in manpower shortage in all technical skills areas [(1)]. For example, high and very high skills gaps are predicted to occur in many developed countries including those in North America in engineering and construction starting as early as 2020 [(2)].

2 The Profile of the Generations Making up the Workforce of the Future

The diversity of the workforce has never been as great as it is today. Multiple generations are in play in the current workforce and each has their own unique profile of desires, interests and requirements. A generation can be defined as a group of people born around the same time and raised around the same place that exhibit similar characteristics (such as communication, shopping, and motivation preferences) and experienced similar trends at approximately the same life stage and through similar channels (e.g., online, TV, mobile, etc.). Generation-shaping trends are most influential as people come of age, which means that members of a particular generation will develop and share similar values, beliefs and expectations. It is important to remember that at an individual level, everyone is different and people exhibit a wide range of trends. However, by looking at people through a generational lens, one can begin to understand how to reach, inform or influence certain segments of a population.

The three key trends that shape generations are parenting, technology and economics [(13)]. To be able to engage the workforce of the future, it is helpful to use this "generational lens" to appreciate what it means to become an employer of choice and to maintain this position over time by adjusting strategies to meet the needs of the changing workforce. Employers will need to address the widely divergent characteristics and idiosyncrasies of several generations at the same time, which is new territory for all companies. This is a tremendous challenge that will require significant adjustments to be successful in the not-so-distant future. It begins with understanding the attitudes and requirements of all the generations that will contribute significant numbers to the workforce of the near future.

The generational timeline applicable to North America is described below along with an overall profile of their overall characteristics. The GI generation (born before 1936) has essentially left the workplace and is mentioned here only for the sake of completeness. Referred to as "the greatest generation", the GI generation had the distinct experience of a) growing up during the great depression after the stock market crash in 1929, b) fighting in WWII, 3) experiencing the fear of communism and prohibition, 4) championing the cause of civil rights and e) joining President Franklin Delano Roosevelt's "new deal" to save the civic fabric of the world and rebuild it.

The generations in play in the current workforce are broadly greneralized below. The range in dates that bracket each generation vary a bit from reference to reference

- **<u>Traditionalists or the Silent Generation (born between 1937 and 1945):</u>** Although not strong in numbers, members of this generation are still active in the workplace and will be for a few more years. Traditionalists lived in a pre-civil rights society and went through their formative years during an era of suffocating conformity, but also during the postwar happiness of peace, jobs, suburban living, television, cars, and rock and roll. [(15)] This generation believed in marriage for life and viewed divorce and having children out of wedlock as unacceptable. The Traditionalists women were pre-feminists who stayed at home to raise children and, if they worked, it was in only certain jobs like teaching, nursing or secretarial. Men pledged loyalty to the company they worked for and generally kept a job for life. Traditionalists had a strong sense of values, were disciplined, cautious and self-sacrificing and were avid readers. Retirement for Traditionalists meant sitting in a rocking chair and living out your life in peace close to home, family and friends.

- **Baby Boomers (born between 1946 and 1964):** Boomers, one of the largest generations in history with about 77 million, defined the American workplace for many years with their workaholic tendencies and work-to-live attitudes. Boomers were optimistic, driven, and team-oriented and tended to be positive about authority, hierarchal structure and tradition. Boomers ushered in the free love and societal, non-violent protests (which did trigger violence). Even though mothers of the Boomer generation were generally housewives, responsible for all child rearing, Boomer women began working outside the home in record numbers, thereby changing the entire nation. This generation was one of firsts: the first generation to have their own children raised in a two-income household where mom was not omnipresent; the first generation with television; the first generation where divorce was beginning to be accepted as a tolerable reality; the first to begin to accept homosexuals; and the first generation to use the word "retirement" to mean being able to enjoy life after the children have left home. Instead of sitting in a rocking chair, Boomer retirees go skydiving, exercise and take up hobbies, which increases their longevity. Boomers are still in the workforce in substantial numbers and their aging has changed and will continue to change America almost incomprehensibly [(15)]. Boomers embrace the notion that "physiology is more important than chronology" [(17)].

- **Generation X (born between 1965 and 1976/1980):** The Gen X generation was raised in the transition phase of written based knowledge to digital based knowledge. Most remember being in school without computers or saw the introduction of computers in middle school or high school. They tend to commit to self rather than an organization or a specific career. This generation averages 7 career changes in their lifetime. It is not normal for a Gen Xer to work for a company for life, unlike previous generations. Gen Xers were raised by the career and money conscious Boomers amidst the societal disappointment over governmental authority and the Vietnam War. They are survivors as individuals, suspicious of all organizations and wary of commitment. Gen Xers were late to marry and quick to divorce [(15)].

- **Millennials or Generation Y (born between 1977 and 1993/96):** This generation is said to be characterized by a short attention span, an instant gratification focus and are technological savvy. They exhibit a sense of entitlement born of the parenting style they were raised under. The generation-defining moment for Millennials was the events of September 11, 2001 ("9/11"). Millennials have very different attitudes toward employment, sales and marketing and these attitudes are challenging many conventional strategies and approaches. Millennials changed every other generation's dependence on social media, text messaging and digital downloading [(14)]. Currently, Millennials have become the largest generation in terms of numbers in the workplace and are the fastest growing generation of customers in the marketplace [(13)]. By 2025, Millennials will make up 75% of the workforce [(13)]. To attract and retain Millennials, employers will be required to make significant changes and investment and provide: state of the art equipment; mobile access and mobile learning (m-learning); workplace flexibility (which is valued over pay) with a teamwork focus; continuous learning opportunities that focus on leadership, team management, self-efficacy building and public speaking [(13)]; ongoing and immediate assessment and feedback on performance; and meaningful mentoring and coaching.

In short, employer-provided technology must meet the expectations of Millennials to enhance retention. Adjustments will need to be made in digital information security and external communication policies as well as cyber security systems to allow the free and open access desired by Millennials.

- **<u>Centennials or Gen Z or iGen (born in 1994/96 or later)</u>**: The start of Gen Z in the North America is closely related to the events of 9/11 in that members of this generation, unlike Millennials, cannot process the significance of 9/11. Members of this generation cannot recall a time when the intranet and mobile devices did not play such a dominant role in society. Gen Z will determine how generations that follow them will ultimately use technology such as wearable devices and the dependence on the "cloud". Gen Z has assumed the Millennial's mantle of the generation to know, understand and engage [(14)]. Gen Z is the first generation to grow up watching their parents (Millennials or Gen X) glued to their smartphones, which drives their desire to have their own smartphone at an early age. Gen Z children have never known a world without computers and cell phones. They tend to leave their conventional toys behind at a younger and younger age, begin to play with computers at ages 4 or 5 and prefer electronics, such as cell phones and video games, to play with. Gen Z members are currently the majority in college and the trendsetters, particularly in technology. This generation will experience a change in the American melting pot. For example, of the record number of births in the US in 2006, 49% were Hispanic. Gen Z has outnumbered the start of the Boomer generation and will easily be a larger generation than the Boomers. Employers must start now to understand and unlock the power of this entirely new generation [(14)] by identifying and addressing their specific needs. Some of the characteristics that define Gen Z include access to mobile devices at an early age (and have their own mobile phone as young as 13).

3 Working in a Multi-generational Workplace

For the first time in the history of North America, we now have four different and distinctive generations working side-by-side; namely, Boomers, Gen X, Millennials and Gen Z [(16)]. For much of the 20th century in North America, (before the year 2000 or so), the workplace was characterized by three generations working side-by-side. During this period, the older workers were generally the bosses and the younger workers toiled under the guidance of an older boss doing what they were asked to do. This workplace was quite structured and most workers followed the established rules and norms of the workplace, both written and implied. Computer technology started to gain a foothold in the early 70's before which time electronic aids and communication devices were largely unavailable. The pace of change was slow until the last 25 years or so.

Roles in the near future will no longer follow the norms and rules of the past. The differences among these generations have started to exert pressures and changes on the workplace and will continue to do at an increased pace in the near future. Generational differences affect all aspects of the workplace including HR processes of recruitment and engagement and remuneration; bonus schemes and timing; working environment; approaches to work; communications both internal and external to the organization; training and development; organizational structure; employee motivation and advancement. The workplace will be changed forever at an unprecedented rate, in many directions and in ways that we cannot imagine today.

One of the major challenges of the workplace in the near future will be communications since it has been shown that how people communicate is based in large part on their generational backgrounds [(16)]. To be successful, companies will have to implement communications systems that will allow the different generations to communication effectively to avoid unnecessary discourse, misunderstandings and confrontation. Companies will also need to offer flexibility in working times, schedules and places as well as the structure of offices. Even today, some companies have already significantly changed the design of their workplace by introducing an "open office" concept when employees work from the office. Another critical area will be the delivery of employee development programs and plans that offer all styles of high quality, professional and technical training as well as employee assessment and meaningful feedback. To be attractive, these programs must appeal to the advanced technology and communication skills of the younger generations. In addition, programs need to reflect the learning audience and the adoption of adult learning methodologies by covering concepts coupled with simulation and on the job real world applications. To add to the complexity this training needs to be conducted in about half the time that was historically taken to develop technical personnel. By the mid 2020's our Gen Xers will also be departing the industry in large numbers such that by 2015, Millennials will comprise the majority of the workforce [(18)].

Companies are well advised to start to profile their workforce in terms of the numbers of employees in each of the generations and to update this profile on a regular basis to fully appreciate the dynamics at play within the organization. Free and open discussions in group settings will help each group to understand these dynamics and discuss the issues of concern and to keep this dialogue open and ongoing. Creativity and innovation are hallmarks of successful companies and many have already made significant changes to foster these traits by effectively managing their intergenerational groups.

4 The Looming Manpower Shortage in Technical Skills Areas

In addition to the challenges related to managing the complex dynamic of a multi-generational workforce involving four distinctive and different generations, a separate challenge is looming in the very near future. This challenge relates to the availability of employees with the requisite skill sets to meet the future staffing demands of companies in the oil and gas industry. The gap between the number of experienced and inexperienced engineers and other technical professionals in the industry will peak in 2016 and is predicted to continue until about 2022. This shortage of experienced engineers is predicted to increase the probability of costly mistakes being made by novices [(18)].

Publications issued by the World Economic Forum in recent years [(1)] [(2)] have highlighted the looming manpower shortage / crisis in all technical skills areas. In the report of the World Economic Forum published in 2010, entitled "Stimulating Economies through Fostering Talent Mobility" [(1)] the disturbing claim is made that *"The global economy is approaching a demographic shock of a scale not observed since the Middle Ages. The working-age populations of many developed economies will start to decline shortly. Numerous organizations will be unable to find enough employees in their home markets to sustain profitability and growth. By 2050, the global population of those 60 years old and older will exceed the number of young (less than 15 years old) people for the first time in history".* Figure 1 below, an exhibit from this report, shows predictions of skilled worker availability for a wide range of industries for the period 2020 to 2030 for a number of countries in the Americas, Europe, Asia and Oceania. The blue color coding indicates how countries are positioned to face shortage challenges with the dark blue indicating a high ability to attract migrant workers because of the attractive attributes of the country while the medium and light blue color code indicates medium and low abilities to attract migrants. The skill gaps are color coded as follows: very high skills gaps (red); high skills gaps (orange); and low skills gaps (yellow) and no skills gaps (white).

Focusing on North America, high and very high skills gaps are predicted to occur in Canada and the US in a number of critical fields including engineering and construction, education and IT between 2020 and 2030. This situation has the potential to curtail economic growth around the world and fundamentally change the workforce challenges. Notably 80% of the impending labor shortage will involve skills, not numbers of workers potentially available. Those companies that are successful at attracting and retaining the best and brightest talent from this diminishing skilled labor pool will have a significant economic advantage over the competition.

The ability to attract and retain talent will be dictated, in part, by meeting the needs of new employees, a very important component of which is the growth and development opportunities provided by their employers. Companies need to reflect an awareness of this looming manpower crisis in their Learning and Management Systems not only to stay ahead of the changing needs of new talent but also to reflect the different demands of the various generations that make up the workforce in the near future. For example, to meet the expectations of Millennials, companies must develop achievable development plans that meet the technical requirements of the associated roles along with the talent pipeline. These plans must be readily available to all and also be sustainable, measurable and repeatable while still providing an accelerated and directed path.

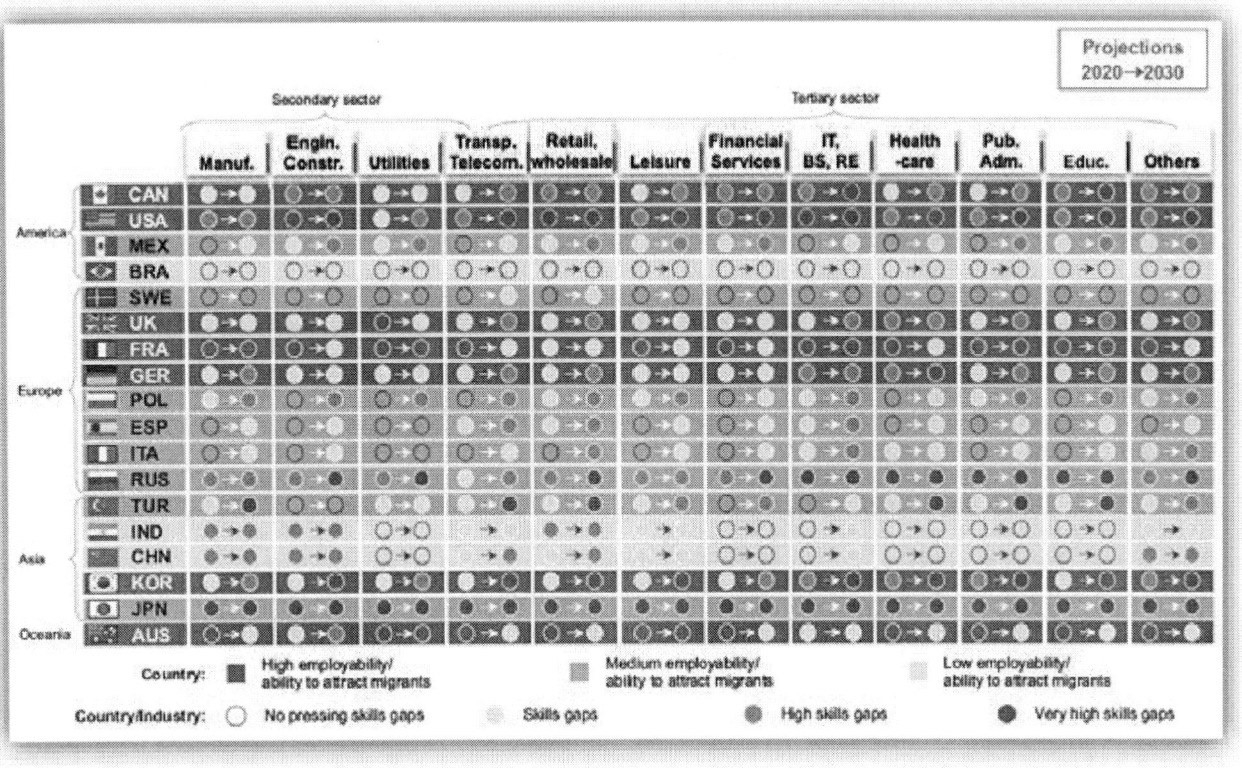

Figure 1 Projections of Skilled Worker Availability 2020 – 2030

4.1 Importance of Skilled Labor Migration

Technical manpower shortages in the past have largely been met by the migration of skilled labor from less developed countries to North America and other desirable workplaces. Figure 2 [(1)] below shows a snapshot of the importance of highly skilled immigrant workers to the economic well-being of a number of countries. The significant share of highly skilled migrant workers as the overall percentage of domestic skilled labor is quite significant as shown in Figure 2 as is the average contribution of highly skilled migrants to the GDP of those countries. Maintaining an increasing the flow of highly skilled migrant workers into North America will be very important to companies in the oil and gas sectors if they are to continue to grow and prosper financially.

Talent mobility can pose short-term challenges to the receiving country in terms of integration as well as sending countries in terms of a "brain drain". In the long term, positive benefits can accrue to both receiving and sending countries. However, these positive benefits could be undermined if protectionism policies are implemented by a number of countries. Such policies can compound the challenges that the oil and gas industry faces in meeting their demands of skilled labor in the very near future. One of the long-term benefits to the sending country [(1)] is the return migration of highly skilled workers to their home countries as conditions and opportunities improve. Planned and mutual cooperation between sending and receiving countries should be developed and implemented to their mutual benefit when the receiving country performs the role of a training ground for skilled workers of the sending country who return on a planned, scheduled basis to repatriate their honed skills. For international companies the needs are largely similar the world over. Career plans, training and development solutions and resources should speak to the global workforce. This drives mobility to get the top talent in the right job at the right time efficiently, consistently and cost effectively to ultimately serve a higher purpose of supporting a desired operating culture.

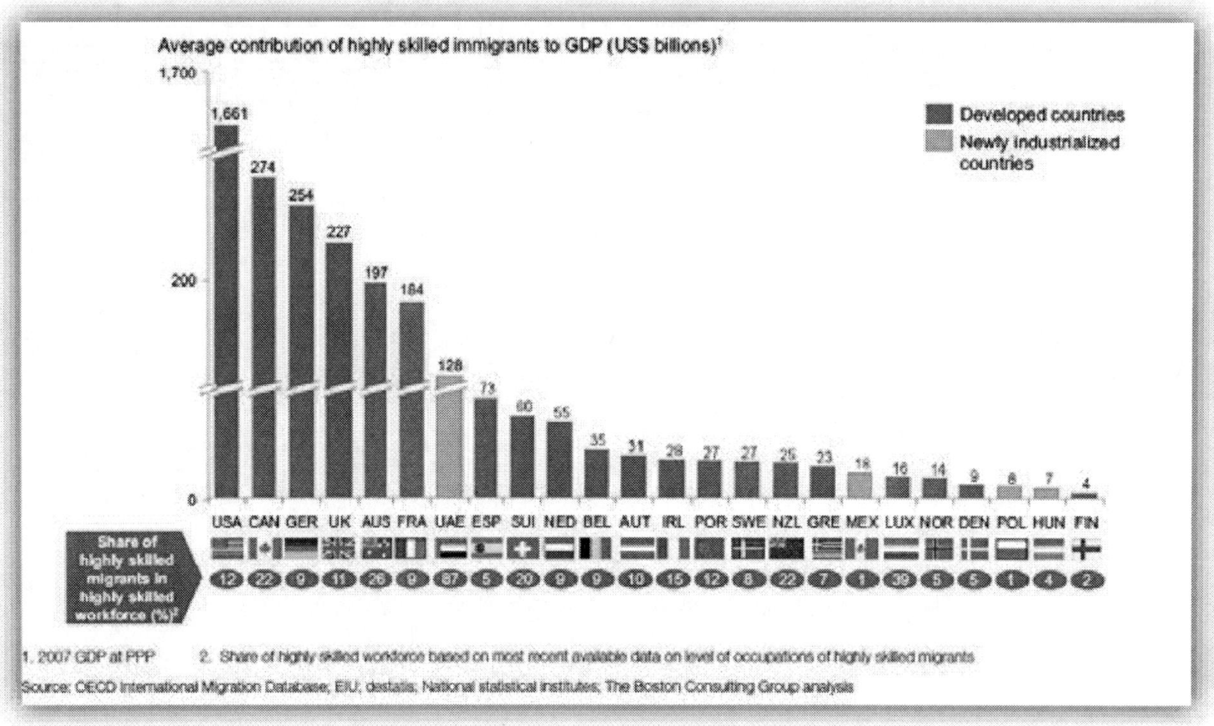

Figure 2 Talent Migration Trends

4.2 Meeting the Global Talent Risk

Figure 3 below, an excerpt from the report published by the in 2011 entitled "Global Talent Risk – Seven Responses" [(2)], summarizes seven key factors that need to be in addressed in order to increase the chances that the future manpower needs of the oil and gas sector in North America are met. Companies can use these factors in developing strategies to frame the approaches they will take to understand the changing workplace environment and to attract, train and retain new technically skilled staff and to retrain and invigorate existing staff to embrace these substantial changes. Those companies that are willing to invest the energy to prepare for this uncertain future will be better positioned to address the looming deficits in skilled talent.

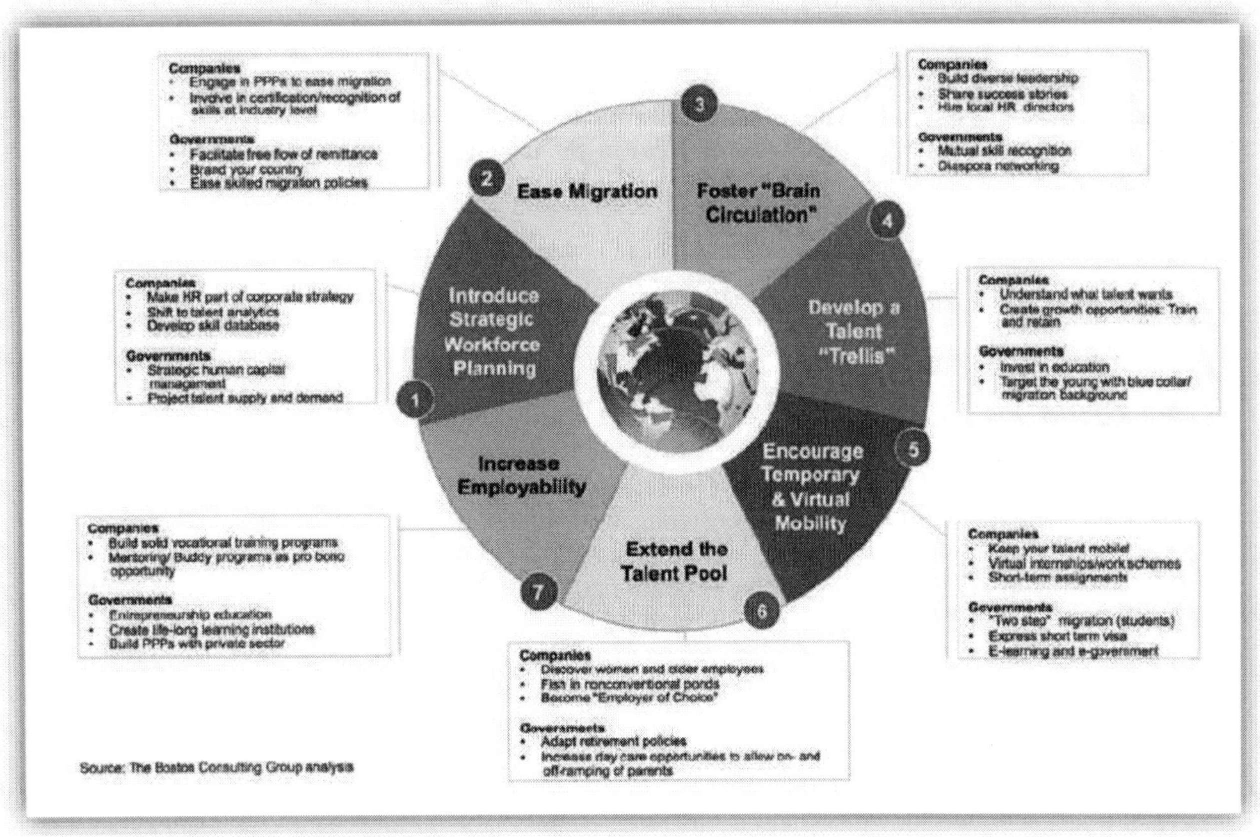

Figure 3 Checklist for Formulating Strategies to Address the Global Talent Shortage Risk

Of the checklist actions outlined in Figure 3, those that are most significant for companies to meet the goal of developing and maintaining a skilled workplace are:

- Build diverse leadership that embraces the HR function as a critical component of corporate strategy, reflects the demographics of the multi-generational workforce at play and fully appreciates and understands what is involved to "become an employer of choice"
- Understanding what their multi-generational staff want in terms of growth opportunities and to cater to these by providing training and retraining suited to each generation including the promotion of virtual internships and work schemes.
- To develop and promote creative ways to extend employment opportunities of older employees in meaningful and unique ways and to encourage the engagement of parents with young children through increased day care opportunities.
- Develop creative and inspiring training programs that embrace the learning attributes of all generations in the workforce and keep these evergreen and exciting and provide ongoing feedback and assessment by way of skills and competency assessments.

Governments have a significant role to play in addressing the looming shortage in skilled labour by supporting company initiatives through investment in education, revised policies related to retirement age and immigration. Only with all parties working in concert will the future demand for skilled labor be met and the negative impacts of skilled labor shortages be avoided.

5 One Company's Approach

In 2014, Occidental Petroleum Corporation (Occidental) identified a demographic and potential competence gap in its skilled labor workforce. Many of its employees were older and approaching retirement age or were younger, newer hires; there were not a lot of employees in between. As older workers retired, there was a risk that the current workforce might lack the experience, skills and numbers to replace them.

To close the gap, Occidental enhanced its training program to focus on technical skills development and competency, using online and in-person training. The company viewed targeted training as a way to retain talent by providing opportunities to promising employees, as well as to facilitate knowledge-sharing by potential retirees within the organization. Key actions focused on training, skills development and competency assessment in all technical areas.

In 2015, the company opened a state-of-the-art training center in Midland, Texas, in the Permian Basin, the base of company's domestic operations. The facility features practical and simulated hands-on training. Operational and field staff had access to customized programs and flexible training schedules, with a focus on providing foundational knowledge and behavioral-based competency. More experienced employees then helped with complementary on-the-job training centered on the "how" and "why" to accelerate the learning process. In the first 12 months of operation, the Midland training center has had approximately 2,300 technical personnel (an average of 190 staff per month) participate in training. The training totaled about 31,000 hours of instruction in a blended environment and involved more than 74 different programs spread over 194 offerings, most of which were integrated with simulation and hands-on activities.

Occidental implemented a new Learning Management System (LMS) in 2016 to support training programs and curricula development with a focus on job role, discipline and position. Employees across the company could now easily access structured training and competency assessments, along with self-directed and just-in-time learning opportunities. The LMS also provided a process and avenue for employees to demonstrate their competence in their current role and qualifications for advancement within the organization.

Beyond the classroom, Occidental is working to create a culture and infrastructure that can support development in a consistent manner inclusive of mentoring programs.

Occidental recently initiated a business transformation program to search for and implement value-added innovative ideas that will propel consistent breakthrough performance. The program seeks to mine and capture the substantial talents and innovative ideas of the entire workforce in a focused and structured manner. The program started in 2016 and will be implemented over the next few years in a multi-phase approach: design, build and deploy, enhance and finally sustain.

All these efforts are facilitated by a mature and integrated technical training team that is well positioned to adjust to and meet the learning styles of all the generations that make up the company's workforce.

6 Conclusions

The workplace in North America is changing very dramatically and within a relatively few years will take on a dynamic never experienced before in history. By 2020, four different and distinctive generations will be working side-by-side. The differences among these generations have started to exert pressures and changes on the workplace and will continue to do so at an increased pace to affect all aspects of the workplace. The workplace will be changed forever in many directions and this change will probably move in directions that are unanticipated today.

To be successful in this new workplace, companies must understand and start to prepare now to meet the challenges that these substantial changes bring. The first step in this process is to understand the detailed profile of the demographics of their current workplace, update this profile on a regular basis, and include an estimate of what the future profile will look like in 3 to 5 years based on the trends of these profiles. These profiles will frame the nature of the multi-generational workplace over time and be the basis of integrated, development plans to meet the future manpower needs of the company.

Critical in this regard is to implement an integrated approach to training and development that:

- reflects the training styles and needs of each generation at play in the workforce
- provides hands-on skills and competency assessment on an ongoing basis
- proposes development programs for highly skilled programs (i.e. risk management, etc.)
- allows cross training and advancement in all directions (the corporate ladder must become a trellis)
- includes a leadership training program that identifies high-potential candidates and provides extensive training and exposure to all aspects of the business,

Working with universities will be important to:

- update technical training program requirements to meet company needs in the short term and long term and actively participate in training, if possible
- develop new programs that speak to the entrepreneurial propensity of the newer generations
- include a substantial number of internships in technical skill areas to promote cooperation and coordination with educational institutions

Active interaction with state and federal governments will be needed to:

- streamline visa processes for foreign students to study at North American universities;
- streamline work visas upon completion of degrees to allow foreign students the opportunity wo work for companies in North America
- develop innovative child care programs that facilitate younger mothers to stay in the workplace;
- revise retirement age restriction where they exist to encourage older workers to keep working after "retirement" age on a basis defined by them

Human Resources departments need to become human capital management departments and an integral component of corporate strategic planning. They should provide training programs that are flexible to encourage retention of staff from all generations. Innovative approaches are needed.

The successful companies of the future will be those companies of today that recognize the nature of the significant changes on the horizon and aggressively put in place plans to embrace the excitement of this new changing dynamic to meet its challenges. Are you ready for the future?

7 References

1. "Stimulating Economies through Fostering Talent Mobility", World Economic Forum in 2010.
2. "Global Talent Risk – Seven Responses", World Economic Forum in 2011
3. "The Coming Neurological Epidemic", Gregory Petsko, Arthur J. Mahon Professor of Neuroscience, Brain and Mind Research Institute, Weill Cornell Medical College, TED talk, February 2008.
4. "How Great Leaders inspire action", TED Talk by Simon Sinek, filmed Sept 2009
5. "The Myth of the Teen Brain", Robert Epstein, Scientific American, April / May 2007
6. "The Amazing Teen Brain", Jay N. Giedd, Scientific American, June 2015
7. "The mysterious workings of the adolescence brain" Sarah-Jayne Blakemore, TED talk, filmed June 2012
8. "The Learning Brain, Lessons for Education", Sarah Jayne Blakemore and Uta Frith, Blackwell Publishing, 2005
9. "The Neurons That Shaped Civilization", Vilayanur Ramachandran, TED India Talk, filmed November 2009.
10. "Growing Evidence of Brain Plasticity", Michael Merzenich, TED Talk, filmed February 2004
11. "Three Clues to Understanding Your Brain", Vilayanur Ramachandran, TED Talk, filmed March 2007
12. "L&D Meets OMG, Learning and Development for Millennials" Bridge, 2015
13. "Generational Breakdown: Info About All of the Generations", Center for Generational Kinetics, Austin, Texas (info@GenHQ.com)
14. "Gen Z Tech Disruption" 2016 National Study on Technology and the Generation after Millennials", The Center for Generational Kinetics and Jason Dorsey, 2016
15. "The Six Generations in America" Dr. Jill Novak, University of Phoenix, Texas A&M University
16. "Mixing and Managing Four Generations of Employees", Greg Hammill, Executive Director for Human Resources Management Studies, Fairleigh Dickinson University, 2005
17. "Making 80 the new 60", Occidental Oil and Gas, internal presentation, Dr. Stewart W. Behie, 2016
18. "Attracting, Retaining & Developing Technical Talent", Matthew Henwood, Technical Training Manager- Growth & Operations Support, Occidental Technical Training Center, September 15, 2016, Marcus Evans Energy Series Conference, Houston.
19. "The Great Generational Shift: The Emerging Post-Boomer Workplace", Bruce Tulgan, Rainmaker Thinking, White Paper, 2016
20. "2016 Trends in the Workplace", CallisonRTKL, 2016
21. "Understanding Generational Differences", Grant Wood Area Education Agency, 2012
22. "Generational Diversity, Teaching and Learning Approaches", Susan A. Johnson and Mary L. Romanello, Nurse Educator, Vol.30, No. 5,September / October 2005
23. "Managing People from 5 Generations" Rebecca Knight, Harvard Business Review, September 25, 2014
24. "Meeting the challenges and opportunities of today's new multi-generational workforce", Fairsail, July 2016
25. "Understanding and managing the 4 Generations in the Workplace", Robert Tanner, Management is a Journey, June 28, 2916

Sustainable Safety – A Case Comprehension

Kumar (Chris) Israni
Principal Consultant - ERM
840 West Sam Houston Parkway North, Suite 600
Houston, TX 77024
Chris.Israni@erm.com

Liam Doyle, CSP
Process Safety Consultant - ERM
3838 North Causeway Boulevard, Suite 3000
Metairie, LA 70002
Liam.Doyle@erm.com

[Insert here any copyright claims you wish to assert]

Prepared for Presentation at
American Institute of Chemical Engineers
2017 Spring Meeting and 13th Global Congress on Process Safety
San Antonio, TX
March 26 – 29, 2017

AIChE shall not be responsible for statements or opinions contained
in papers or printed in its publications

Sustainable Safety – A Case Comprehension

Kumar (Chris) Israni
Principal Consultant - ERM
840 West Sam Houston Parkway North, Suite 600
Houston, TX 77024
Chris.Israni@erm.com

Liam Doyle, CSP
Process Safety Consultant - ERM
3838 North Causeway Boulevard, Suite 3000
Metairie, LA 70002
Liam.Doyle@erm.com

Keywords: Process Safety, Sustainable Safety, Risk Management, Human Factors, Asset Integrity, Accelerated Learning.

Abstract

We have seen a lot of changes in the approach to process safety and risk management over the last four decades, within the Oil and Gas (O&G) and Chemical sectors. Learnings from catastrophic incidents in the industry, led to the evolution and enhancement of risk techniques, as well as global regulatory changes. The last decade has also seen the evolution of the concept of sustainability. One of the key elements for a company in evaluating their sustainability performance is safety; thus tying safety into each company's overall goal to strive towards operation excellence.

This paper elaborates on the sustainable and sequential application of process safety methods in a risk based manner for the entire lifecycle of the facilities. That assures longevity for the facilities and thus operational excellence. The case comprehensions discussed in this paper will highlight the ways we've witnessed/assisted various groups and organizations to achieve success in this area.

1 Introduction

Effective and sustainable process safety programs are vital to the overall health and viability of a company from a business continuity perspective; as evidenced by the devastating business impacts previously incurred by those companies that failed to effectively sustain such programs. Unfortunately, in our experience, we find that even though some companies may spend up to 25% of their operational expenditures on safe production programs, only about 10% of that spending generally ends up going towards programs/activities capable of driving actual improvements in safety and operational performance [3,8]. Sometimes this may be due to inefficient or unwieldly leadership programs. Other times it may be the result of an unbalanced allotment of resources aimed more towards controlling process hazards via personal, compliance-based safety programs, instead of mitigating and/or eliminating process hazards, efficiencies, and/or weak spots via process safety and risk management programs. However, regardless of the cause, such investments are highly inefficient and simply unsustainable. Thus, we recommend such companies redirect their efforts more towards those programs actually capable of driving sustained improvements in process safety and operational performance, such as Sustainable Safety [8].

2 Safety Leadership Programs

Several leadership programs currently in use today focus on encouraging workers to "own" safety as a common practice. Some of the methods used by such leadership programs to encourage workers to "own" safety include: (1) empowering employees to take action when they witness unsafe behaviors; and (2) educating employees about the off duty activities that could possible lead to unsafe acts at work. However, in order for these programs to be effective the management teams facilitating the programs must first develop a work environment that's consistent with the different rationales and expectations present within their multi-cultural workforce [4].

For example, as we've discussed in a previous publication [4], failing to account for the regional customs in the workforce can inhibit the very feedback such leadership programs are desperate to receive. In this publication [4] our colleague, Tianherng (Joshua) Yang, provides an example of a workforce hesitant to report the major safety faux pas being noticed amongst peers and supervisors to the appropriate authorities. In most cases, this hesitation would likely be attributed to a fear of backlash and thus addressed by promoting "whistleblower" protection programs. However, in this workplace, the workforce culture placed an exceptionally high emphasis on respect. Thus, the root cause of this hesitation was also due, in part, to the workforce's fear of disrespecting their peers and supervisors [4]. Failing to account for the respect-based aspects of this workforce's culture would have prevented any employee feedback encouragement programs, such as a "whistleblower" protection program, from achieving its intended objective.

3 What is Sustainable Safety? [3]

Sustainable safety is a concept we've termed to describe how the people working within a typical manufacturing facility utilize the facility's existing processes (management systems, procedures, permits) to operate and maintain the plant (equipment, machines, vehicles, etc.) aspect of the facility's operations. As shown in the figure below, exceptional operational performance typically occurs at the interface between the People, Process, and Plant elements; with Technical Safety (Process Safety/Risk Management) generally occurring at the interface between the Process and Plant elements.

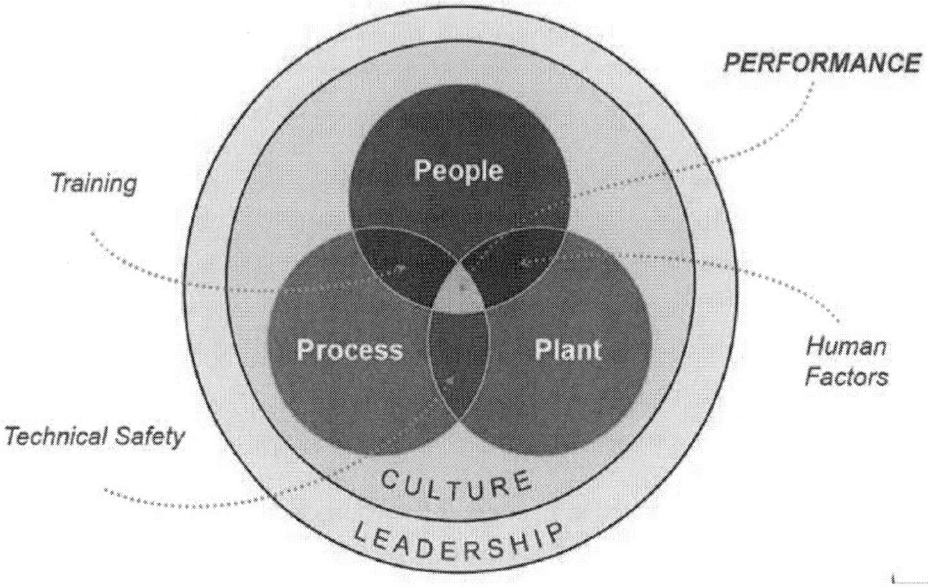

For example, written operating procedures (Process), developed to increase operator productivity, may fail to achieve its intended goal if designed without the input of those (People) by which it's intended to be used. Additionally, such procedures may also fail to realize their intended objective if they're designed without accounting for recent changes made to process equipment (Plant). Consequently, operational performance is generally tied to how well each element interacts with one another.

For the purposes of this paper we have chosen to highlight how we've seen some companies incorporate their existing technical safety management systems into Sustainable Safety to drive considerable improvements in operational performance, risk reduction, and facility longevity.

4 How Does Technical Safety Fit into Sustainable Safety – Case Comprehensions?

In our experience, the best way to promote sustainable safety, from a process, or "technical safety" standpoint, is through the development and implementation of focused, risk-based approaches. Such approaches, when implemented properly, can have significant returns, as further discussed in the examples outlined below pertaining to standard operating procedures, asset integrity, and human factors.

3.1 Standard Operating Procedures

As stated by the Center for Chemical Process Safety (CCPS) in their book, "Guidelines for Risk Based Process Safety" [2], a consistent high level of human performance is a critical aspect of any process safety program. Indeed, a less than adequate level of human performance will adversely impact all aspects of operations. Without written procedures, a facility can have no assurance that an individual operator will consistently execute a particular task in its intended manner [2]; thus leading to an increase in operational waste and a decrease in operational efficiency due to operator error/ignorance.

In 29 CFR 1910.119(f), OSHA outlines its regulatory requirements for standard operating procedures. Such requirements places an emphasis on ensuring employers provide employees with clear, concise, and most importantly, accurate, work instructions for safely conducting their assigned tasks in, on, or near the employer's PSM-covered process. These requirements were put in place, in part, because of OSHA's view that incidents are less likely to occur if written operating procedures are developed so that even a new employee, or someone relatively inexperienced, may perform a task and/or respond to a given event in a pre-considered and prescribed manner [6].

Interacting with many key players in the industry, we have observed that when companies develop and implement new/revised standard operating procedures in a manner that effectively accounts for the "People" and "Process" aspects of Sustainable Safety, they typically realized a 7-10% gain in process productivity, as well as a significant reduction in operational waste because of fewer operator errors and rework [3]. In fact, it's been said that in addition to helping ensure process safety, procedures can also help ensure reliable human performance, which in turn helps ensure predictable product output, yield, and quality [2]. Likewise, less than adequate procedures can result in less than adequate human performance; which can have adverse effects on all aspects of a facility's operation [2].

In our experience, such gains are typically realized when standard operating procedures are developed/implemented in conjunction with effective training programs [3]. Indeed, using operating procedures as a training aid can help: (1) familiarize trainees with the content and format of procedures, (2) reinforce the value of the procedures, and (3) allow trainers or other subject matter experts to identify errors or omissions in procedures [2].

The current trend in the industry, is to use an accelerative learning (AL) approach to training [7]. Enhancement of training programs, using the AL techniques has previously rendered a 30% gain in knowledge retention and on the job implementation of the concepts that are rolled out.

An example of this includes a recent project for an injection molding manufacturing process in which we overhauled the facility's operating procedures and developed a structured on the job training program to help the facility qualify its operators. By incorporating the facility's updated procedures into its revised operator training program, the facility has helped ensure the future accuracy and ongoing use of its written operating procedures [3, 7].

3.2 Asset Integrity

To paraphrase the CCPS [2], asset integrity programs generally consists of a set of written standards, procedures, policies, and related training activities meant to help assure process equipment is fabricated, installed, and maintained in a proper manner. Such programs typically utilize a set of standards, or technical bases, such as recognized and generally accepted good engineering practices (RAGAGEP), to develop/define inspection, testing, and preventative maintenance (ITPM) activities for various types of process equipment. In our experience, the most effective programs typically also utilize a risk-based approach to ITPM activities that adopts a "living" inspection regime based on the actual operating and environmental conditions at the facility, as well as the condition of the facility's equipment when last inspected. Thus, placing an emphasis on ensuring the ITPM activities tied to the areas of greatest concern are always completed.

In fact, the CCPS has indicated that safe operations are directly and closely correlated with reliable operations, and conditions that lead to safe and reliable operations generally lead to high overall plant availability and other business related benefits [2]. Moreover, our experience has shown that implementing a robust asset integrity program can help facilities increase equipment "up-time", extend the useful life of their equipment, and achieve an overall increase in productivity; thus, reducing overall capital expenditures.

For example, we previously implemented an asset integrity and reliability program for a company that lead to an approximate 18% increase in equipment "up-time", as well as a corresponding increase in overall productivity. This implementation included the development of an enhanced mechanical integrity program that better aligned the facility's existing ITPM activities with applicable RAGAGEP [3].

3.3 Human Factors

The term Human Factors represents a concept based on the study of operators, managers, maintenance personnel, and others in the work environment, and of the factors that generally influence humans in their relationship with a technical installation. Thus, human factors related programs and activities generally involve activities meant to help assure machines, operations, and work environments are designed so that they match human capabilities, limitations, and needs [1]. In terms of Sustainable Safety, human factors activities are best represented at the interface between the "People" and "Plant" aspects of a company's operation. Thus, the success of a company's Human Factors program is directly influenced by how well it's able to incorporate the "People" and "Plant" aspects of its operation into its human factors related programs/activities.

For example, a non-regulated OSHA PSM cotton processing pilot plant facility in the eastern United States had recently experienced multiple fire events. The site had previously completed traditional HAZOPs for each of its processes to prevent such events from occurring, but had failed to fully address the human-machine interface of its process. Thus, the facility asked us to help facilitate a critical control study of the site's processes to identify the gaps and disconnections existing in its human factors dependent processes. The resulting corrective actions have since been said to have helped increase operator efficiency and substantially minimize the site's overall risk of fire [3].

5 Conclusion

Effective and sustainable process safety programs are vital to the overall health and viability of a company from a business continuity perspective; as evidenced by the devastating business impacts previously incurred by those companies that failed to effectively sustain such programs. Thus, we feel that it is vital that companies focus much more of their safety/risk reduction-related efforts towards those programs capable of driving sustained improvements in process safety and operational performance, as well as overall facility longevity, such as Sustainable Safety.

6 References (examples given to show format).

(1) CCPS. *Guidelines for Hazard Evaluation Procedures, 3rd Edition.* Center for Chemical Process Safety, American Institute of Chemical Engineers, New York, NY, 2008.

(2) CCPS. *Guidelines for Risk Based Process Safety.* Center for Chemical Process Safety, American Institute of Chemical Engineers, New York, NY, 2007.

(3) "Achieving Operational Excellence Through A Sustainable Approach Within Process Safety", Kumar (Chris) Israni, Steve Hawkins, Presented at GCPS 2016.

(4) "Safety Leadership and Implementation- Independent of Culture", Kumar (Chris) Israni, Sarah Acton, and Tianherng (Joshua) Yang, Presented at GCPS 2014.

(5) "Safety Culture and Leadership - Independent of Language", Kumar (Chris) Israni, Sarah Acton, Yaneira Saud and Sandra Lovell, Presented at Mary Kay 2016.

(6) Occupational Safety and Health Administration. "Process Safety Management of Highly Hazardous Chemicals." U.S. Department of Labor, Occupational Safety and Health Administration. 1910.119

(7) "Applying the "Concept of Accelerative Learning" for Design and Delivery of Process and Personnel Safety Leadership Programs in Oil & Gas Assets", Kumar (Chris) Israni and Collen Watson, Presented at SPE – HSSE New Orleans 2017.

(8) "Achieving Sustainable Safety" Environmental Resources Management (ERM), et al. Internal ERM document, 02-01-2017.

479491 Are Our Current Safety Challenges Different Than 1974 or 1984? Have Wereally Learnt from Accidents? Where Are the Gaps, What Are We Missingand Why? Learn What We Have Not Learnt in Last 40 Years, Keptrepeating the Same Mistakes

Abstract:

Despite the extensive development in technology, a considerable reduction in process safety accidents was expected, but slight improvements are realized. If we look back 50 years, no doubt, the likelihood and severity of consequences for majority of accidents have reduced, however the root causes are still unchanged. So, it is worth debating that why do we keep repeating similar accidents in spite of using latest tools, techniques and learning.

The presentation will cover what has changed and what not over last 50 years. The author will discuss the practical examples about most common gaps, current safety challenges, and most needed process safety improvements to eliminate the learning gaps and prevent the recurrence of accidents. The role of good safety culture and information flow between management and shop floor will also be discussed in relevance to accident prevention.

Several methods and techniques will be presented to identify the opportunities for improvement and tips for effective implementation during design and operating stage of the facility. Emphasis will be given on prevention than mitigation by applying hierarchy of control.

Presentation will conclude by explaining how a robust management system, committed management and motivated employees play a vital role in preventing the recurrence of accidents by eliminating the root causes.

Copyright © American Institute of Chemical Engineers. All rights reserved.

481499 Life Cycle Analysis of Hydrogen Gas Used for Biogas Upgrading Process in Expanded Granular Sludge Bed Reactor

Abstract:

<align="center">Life Cycle Analysis of Hydrogen Gas Used For Biogas Upgrading Process in Expanded Granular Sludge Bed Reactor

Shruti S. Karambelkar, Haider Al-Rubaye, Manohar M. Shivashankaraiah, Joseph D. Smith, PhD.

Chemical & Biochemical Engineering Department, Missouri University of Science and Technology, Rolla, USA

Abstract

The high demand for energy around the world and the environmental effects resulting from the use of the fossil fuels, makes many industrial sectors focus on the alternative energy. Biogas Production from anaerobic digestion forms an essential part of alternative and renewable energy development.

The biogas mainly consists of methane, carbon dioxide and trace amounts of hydrogen sulfide and ammonia. Hydrogen gas produced in the first two steps of the anaerobic digestion - Hydrolysis and Acidogensis is released to the atmosphere during the process. Injecting the same amount of released Hydrogen into the last two steps of the Anaerobic Digestion - Acetogensis and methanogenesis, improves Methane composition in the biogas, up to 88%. On injecting Hydrogen, it reacts with the produced carbon dioxide to produce more methane, thus, enhancing methane in the biogas. This increase in the methane composition will be accompanied by a high production rate since the specific volume of the methane is four times more than hydrogen and the volumetric energy content for methane is higher (40 KJ/L) than that of hydrogen (12.78 KJ/L).

Life Cycle analysis has been conducted to check the feasibility of using Hydrogen gas to upgrade the biogas so it could be accounted as a natural gas and used in all natural gas equipment.

Copyright © American Institute of Chemical Engineers. All rights reserved.

478350 Permylene™ Membrane Technology By Imtex Membranes Corp. - the Cost Effective Alternative to Distillation for Olefin Purification and Extraction

Abstract:

Imtex Membranes Corp. is a leading innovator of advanced membrane separation technology currently focusing on olefin/paraffin separation for application in the petrochemical and refining industries. Following years of research and development, Imtex currently possesses the only high performance, commercially viable alternative to traditional distillation techniques for olefin purification and extraction - Permylene™. The technology's potential for substantial economic and environmental benefits is expected to set new standards across the industry.

For over 30 years the petrochemical industry has been searching for commercially viable alternatives for olefin extraction and purification. The capital and operating costs associated with conventional separation via distillation have hindered project returns and ongoing operational profitability.

Imtex has now developed such an alternative that significantly reduces upfront capital investment and ongoing operational costs. Using proprietary membrane technology, this system is also able to perform separations that have to date been deemed impractical or too costly to pursue. This revolutionary breakthrough is seen as the next advancement in olefins processing technology, offering a high performance alternative to the status quo. Imtex is currently working with several industry leaders to deploy demonstration systems in their petrochemical facilities.

Imtex continues to develop membrane separation technology for a range of applications involving C_2, C_3 and C_4 olefins. Imtex is actively searching for additional demonstration opportunities within the industry, which is the first step towards users realizing the significant benefits the technology will deliver.

Imtex will present the characteristics, attributes and benefits of the Permylene technology, and will illustrate several comparative cases that demonstrate the commercial investment and operational cost advantages of the technology versus conventional olefin/paraffin separation technologies.

Copyright © American Institute of Chemical Engineers. All rights reserved.

Hydrogen Storage and Transportation using SPERA Hydrogen Process for realizing Low-Carbon Society

Yusuke Nakajima, Hironori Kawai and Tomoyuki Mikuriya
Chiyoda Corporation, Yokohama, Japan

Introduction

Hydrogen is a clean energy produced from any form of primary energies. It is expected that hydrogen will promote the utilization of renewable energies and solve the problems related to CO_2 emissions.

The establishment of storage and transportation technologies, similar to those used in the petroleum or gas industries, is required for the global utilization of hydrogen.

Chiyoda Corporation has developed the organic chemical hydride (OCH) method using a toluene (C_7H_8) / methylcyclohexane (MCH: C_7H_{14}) system as the basis for the large-scale storage and transportation of hydrogen energy, and demonstration tests were successfully carried out to verify its expected performance. We call the MCH in this system "SPERA (Latin word for hope) Hydrogen", which reflects our ambition to establish a hydrogen supply chain.

At the AIChE Spring Meeting in 2014, we already reported the OCH method and its development status[1)]. This paper introduces the hydrogen supply chain concept, results of the demonstration tests, current activities and future plans of the SPERA Hydrogen process towards a low carbon society.

1. Hydrogen Supply Chain

Figure 1 shows the hydrogen supply chain concept which was proposed in Japan in 2004, by using a hydrogen storage and transportation system based on the OCH method[2-3)].
The hydrogen supply chain concept will be developed in two steps. In the first stage, hydrogen from fossil resources, called clean hydrogen, is utilized with the combination of carbon dioxide capture and storage (CCS), enhanced oil recovery (EOR) or enhanced coal bed methane (ECBM).

In the second stage, hydrogen from renewable energy (wind power, hydraulic power, solar power etc.) will gradually be introduced in order to realize a low-carbon society.

2. Organic Chemical Hydride Method

The OCH method is a way to store and transport hydrogen under ambient temperature and pressure by using a Liquid Organic Hydrogen Carrier. Hydrogen is fixed to aromatic compounds such as toluene by the hydrogenation reaction, whereby toluene is converted to MCH. Hydrogen is generated by a dehydrogenation reaction from MCH at the hydrogen user site, whereby recovered toluene is recycled back to hydrogen production site. The reaction formula is shown below.

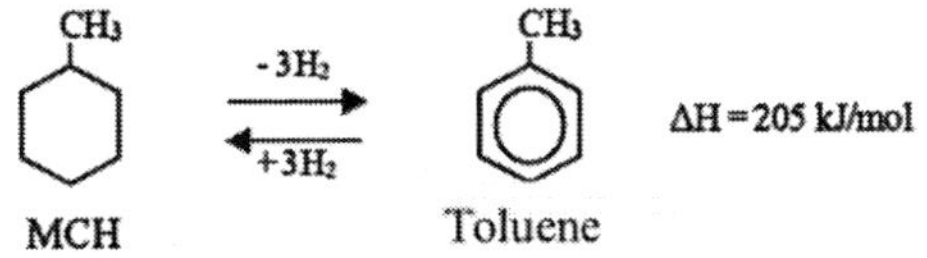

MCH Toluene

Since both toluene and MCH are components of gasoline and are in liquid state under ambient conditions, gaseous hydrogen can be stored in MCH on a large scale and transported in the same way as gasoline.

Table 1 shows the comparison of physical properties of typical organic hydride systems. Chiyoda Corporation has employed the toluene and MCH system, because these components can be kept in a liquid state in the temperature range of -95 to 101℃. Therefore, the system does not require a solvent, unlike other systems which need a solvent to maintain a liquid state.

The OCH method has the following features:

(1) Hydrogen gas volume is reduced to <1/500 of that under ambient conditions

Hydrogen has the lightest molecule and has the largest volume by weight. It is explosive and therefore has potential risk when stored in a gaseous state. In order to reduce the volume to under 1/500 by compression, over 500 bar is required. However, since the OCH method uses chemical reactions, the hydrogen is stable in the MCH liquid, and the gas volume is reduced to < 1/500 under ambient conditions.

(2) Negligible hydrogen loss under long term, large-scale storage

This method does not require high pressures or extremely low temperatures, as with LNG (-163℃) or liquefied hydrogen (-253℃). Toluene and MCH exist as liquid under ambient conditions, therefore, there is no need for boil-off gas treatment.

(3) Utilization of existing infrastructure

Toluene and MCH are components of gasoline and are not corrosive, therefore, conventional gasoline or chemical storage tanks or tankers can be utilized for storage and transportation. This feature decreases infrastructure preparation costs and reduces the risk of large-scale hydrogen use to those associated with gasoline.

3. Dehydrogenation Catalyst

Since the 1980s, it has not been possible to commercialize the MCH dehydrogenation process due to severe catalytic deactivation. Chiyoda Corporation has developed a new dehydrogenation catalyst (SPERA Hydrogen catalyst) and a new dehydrogenation process by scale-up engineering technology.

The SPERA Hydrogen catalyst (S-Pt/Al_2O_3) has uniformly well-dispersed nano-size Pt clusters (< 1nm) on an alumina support . This well-dispersed Pt catalyst enables high reactivity and prevents catalytic deactivation by coking.

Figure 2 shows the result of the catalyst performance test. Over a period exceeding 10,000 hours, the catalyst achieved over 95% MCH conversion, over 99.9% toluene selectivity and over 1,000 $Nm^3/h/m^3$-cat of hydrogen production at a bench facility.

4. SPERA Hydrogen Process

Chiyoda Corporation has developed the new dehydrogenation process using this developed catalyst, and also improved the hydrogenation process based on commercialized technology. Figure 3 and Figure 4 shows simplified process flows of the hydrogenation and dehydrogenation process respectively. The remarkable feature of the SPERA Hydrogen process is its simple process configuration. Each process feedstock reacts in the fixed bed tubular reactors in vapor phase, and high product yield is achieved with simple vapor and liquid separation drums.

Hydrogenation process

The toluene feed is vaporized (Eq. 1) and mixed with hydrogen including recycled gas. The feed mixture is superheated to reaction temperature (Eq. 2) and enters the top of the hydrogenation reactor (Eq. 3), which is a fixed-bed tubular type reactor charged with the hydrogenation catalyst. In the tubes, toluene reacts with hydrogen to produce MCH. Hydrogenation is an exothermic reaction, so in order to control the reaction temperature, the generated heat is removed by a cooling medium, such as a steam generation system.

The effluent gas from the reactor is cooled and the condensed MCH is separated (Eq. 6), with the hydrogen gas being recycled to the reactor. The liquid MCH product is sent to a storage tank.

Dehydrogenation process

The MCH feed is vaporized (Eq. 11), and superheated in a charge heater (Eq. 12) before entering the dehydrogenation reactor (Eq. 13). The reactor is of a fixed bed tubular type charged with the dehydrogenation catalyst in the tubes. Dehydrogenation is an endothermic reaction, so a heating source is required. A hot oil system is typically applied (Eq. 14) to supply this reaction heat.

The effluent gas from the reactor is cooled to separate the hydrogen and toluene (Eq. 16). The separated hydrogen is compressed and sent to users (Eq. 17). Product H_2 treatment (Eq.18) is the optional facility to meet any product hydrogen specifications.

5. Demonstration test of the SPERA Hydrogen Process

Since April 2013, demonstration tests have been conducted in a 50 Nm^3-H_2/h capacity demonstration plant located in Yokohama, Japan. The plant has both hydrogenation and dehydrogenation units, and the conversion between toluene and MCH has been conducted repeatedly. Figure 5 shows this demonstration plant. Although the plant capacity is only 50Nm^3/h-H_2, the catalyst and reactor tubes are the same as those of a commercial plant. Therefore, the process is able to be scaled-up by increasing the number of catalyst tubes.

Figure 6 shows the performance of the hydrogenation and dehydrogenation processes. The upper graph shows operation load and the lower graph shows the performance of each process (hydrogenation and dehydrogenation) performance. The ratio to equilibrium is a ratio of real MCH or toluene conversion to equilibrium conversion. Excluding a period to change operating conditions, we have achieved almost 10,000 hours of successful and trouble-free operation with excellent performance. In addition, the stable recycling of toluene and MCH was also achieved, which has proved that the SPERA Hydrogen process is a feasible technology.

6. Current activities and future plans

6.1 Global hydrogen supply chain demonstration

The next target is to conduct a demonstration of the global hydrogen supply chain, between the hydrogen production site (abroad) and the end user (Japan). In the operations, the produced toluene/MCH are repeatedly transported by shipping, and the transported hydrogen is used in thermal power plants. Through these demonstrations, the actual chain operations are confirmed and then, the global hydrogen supply chain can become real.

6.2 Utilization of renewable energy

To realize a low-carbon society, hydrogen must be produced from renewable energies. There are several technical challenges for harnessing renewable energies. The first is utilizing the surplus energy from wind or solar power, which is generated by the gap between power output and power demand. The second is utilizing unstable power, especially as observed in wind power, which changes from day to day and by the seasons. Furthermore, the renewable energies are unbalanced in global supply and demand. Therefore, the SPERA Hydrogen will be an energy carrier that absorbs this surplus and unstableness of renewable energy and make up any energy imbalances.

In order to introduce renewable energies into in the SPERA Hydrogen process, we have tested and planned the integration of water electrolysis using wind power and SOFC (Solid Oxide Fuel Cell) power generation. The overall scheme of the plan is shown in Fig. 7. A wind power simulation system, alkaline-water electrolyzer and hydrogen purification unit are combined with the existing demonstration plant (described in chapter 5).

The wind power simulation system is applied to an alkaline-water electrolyzer to test the hydrogen generation using unstable renewable energies. This unstable power generates "unstable flow of hydrogen", therefore, a control system that follows the unstable flow will need to be developed. The obtained data will be used to commercial scale-up. The produced hydrogen in the electrolyzer contains impurities (alkaline mist, water, oxygen), so the hydrogen purification unit is applied for the stable operation of the toluene hydrogenation.

The SOFC generates power and surplus heat, the surplus heat can be utilized as heating source of dehydrogenation reaction. The required amount of impurities' removal for the SOFC and the recovery of the surplus heat from the SOFC are studied in laboratory tests. After establishment of the overall integrated system (Fig. 7), variable wind power can be stably supplied using SPERA Hydrogen as a buffer.

6.3 Dehydrogenation facility for hydrogen refueling stations

Considering its ease of handling, the SPERA Hydrogen can be supplied for many purposes. In the future, the demand of hydrogen will increase, not only for industrial purpose, but also for use in logistics, residential and commercial buildings, and so on.

In order to meet this demand, we plan to use the SPERA Hydrogen to supply hydrogen to FCVs (Fuel-Cell Vehicles) by establishing dehydrogenation facilities for hydrogen refueling stations.

Japanese government plans to introduce more than 800,000 FCVs by 2030, and hydrogen refueling stations have started to open and operate in major cities.

As a hydrogen-supply facility for the refueling stations, Chiyoda Corporation has planned an on-site type dehydrogenation facility. As the refueling stations are located in towns, the hydrogen-supply facilities need to be compact, and the produced hydrogen must be purified to meet the specifications for FCVs. As a purification unit, PSA (Pressure Swing Adsorption) is applied and the off-gas from PSA is used for fuel gas as heating source of dehydrogenation.

We plan to start operation of the demonstration plant in Yokohama, Japan in 2017.

7. Conclusion

Chiyoda Corporation proposes a hydrogen supply chain using the SPERA Hydrogen process. The basic technologies were established by achieving around 10,000 hours of successful demonstration operation in Yokohama, Japan. To realize a low-carbon society, we are planning to conduct demonstrations for the global hydrogen supply chain, the utilization of renewable energy and the dehydrogenation facilities for hydrogen refueling stations. These current activities and future plans are subsidized by the New Energy and Industrial Technology Development Organization (NEDO).

References

1) Shiraga M. et al. (2014) AIChE Spring Meeting and Global Congress on Process Safety, 89-C
2) Okada Y. et al. (2004) 15th World Hydrogen Energy Conference, 30B-5
3) Okada Y. et al. (2006) Int. J. Hydrogen Energy, 31, 1348

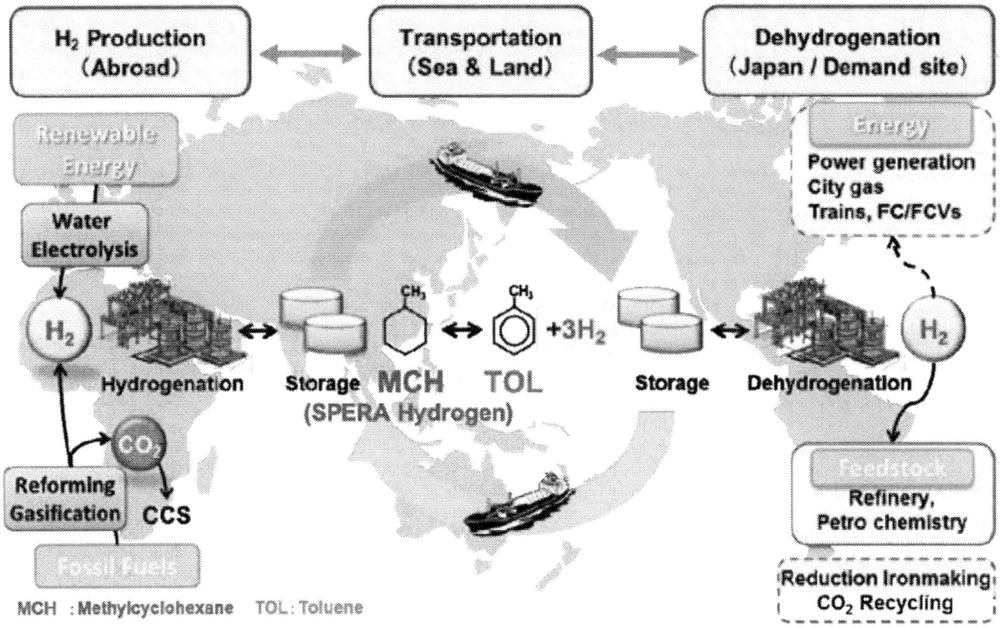

Fig. 1: Hydrogen supply chain concept

Table1: Physical properties of typical organic hydride system

		MCH–Toluene		Cyclohexane–Benzene		Decalin–Naphthalene	
		MCH	Toluene	Cyclohexane	Benzene	Decalin	Naphthalene
Molecular formula		C_7H_{14}	C_7H_8	C_6H_{12}	C_6H_6	$C_{10}H_{18}$	$C_{10}H_8$
Chemical equation		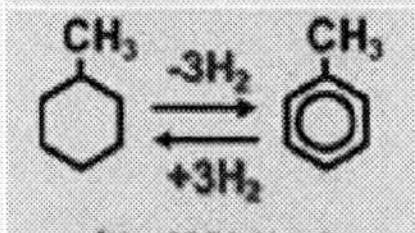 ΔH = 205 kJ/mol		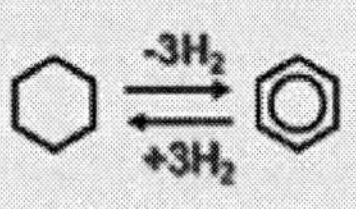 ΔH = 206 kJ/mol		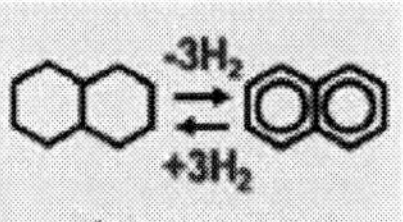 ΔH = 332 kJ/mol	
Molar mass(g/mol)		98.2	92.1	84.2	78.1	138.3	128.2
Phase @room temp.		Liquid	Liquid	Liquid	Liquid	Liquid	**Solid**
Density(g/cm^3)		0.77	0.87	0.78	0.87	0.90	0.98
Melting point (Degrees C)		-127	-95	7	6	Cis:-43 Trans:-30	80
Boiling point (Degrees C)		101	111	81	80	Cis:195 Trans:186	218
H_2 store density	(wt%)	6.2	–	7.2	–	7.3	–
	(kg-H_2/m^3)	47	–	56	–	65	–

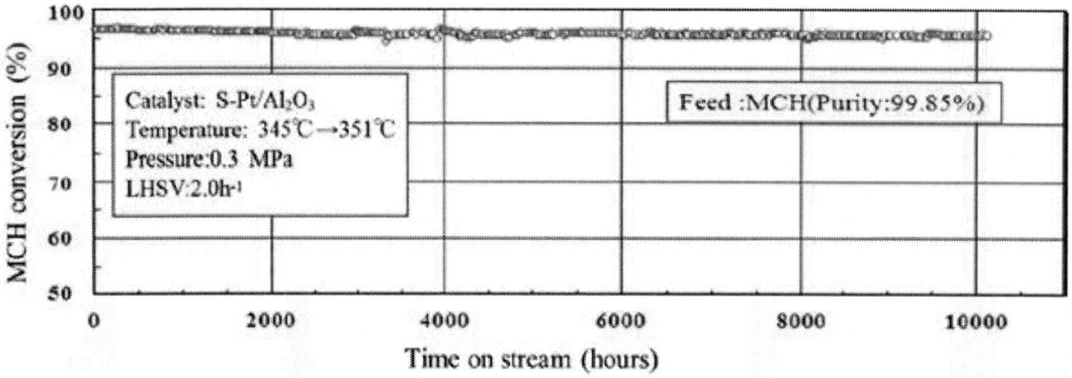

Fig. 2: Result of dehydrogenation catalyst performance test

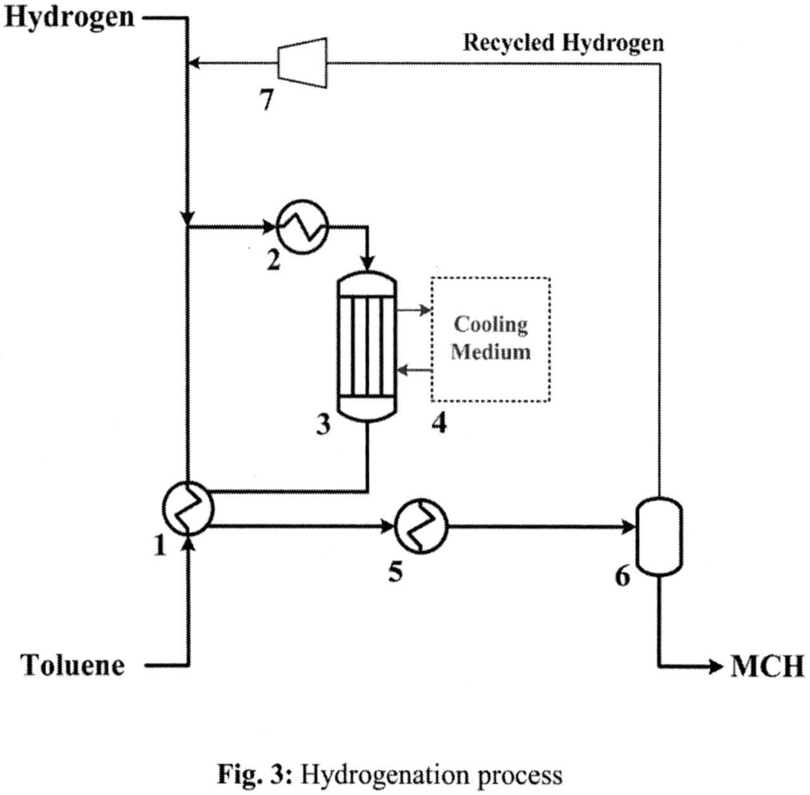

Fig. 3: Hydrogenation process

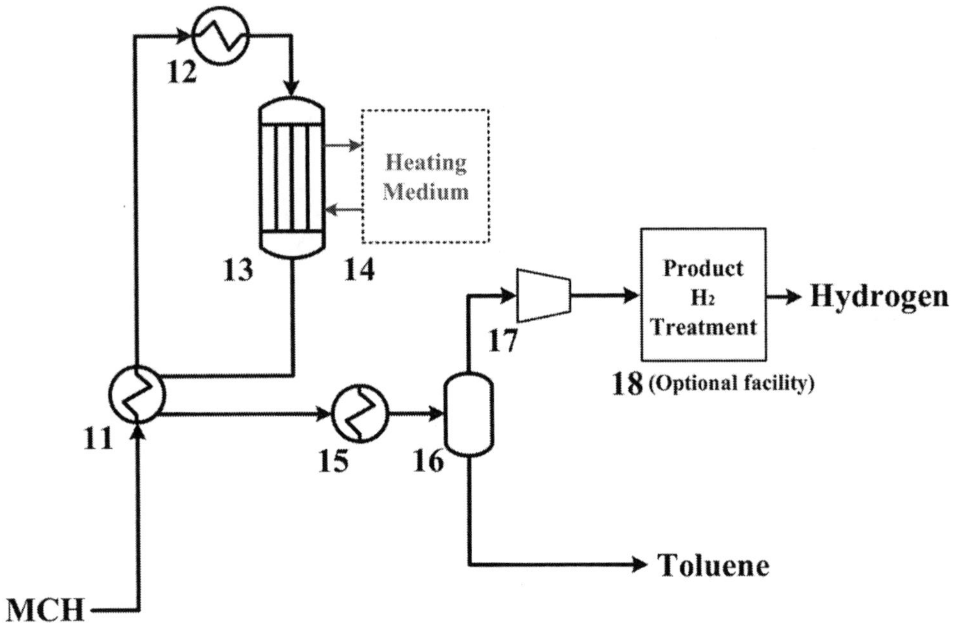

Fig. 4: Dehydrogenation process

Fig. 5: Demonstration plant

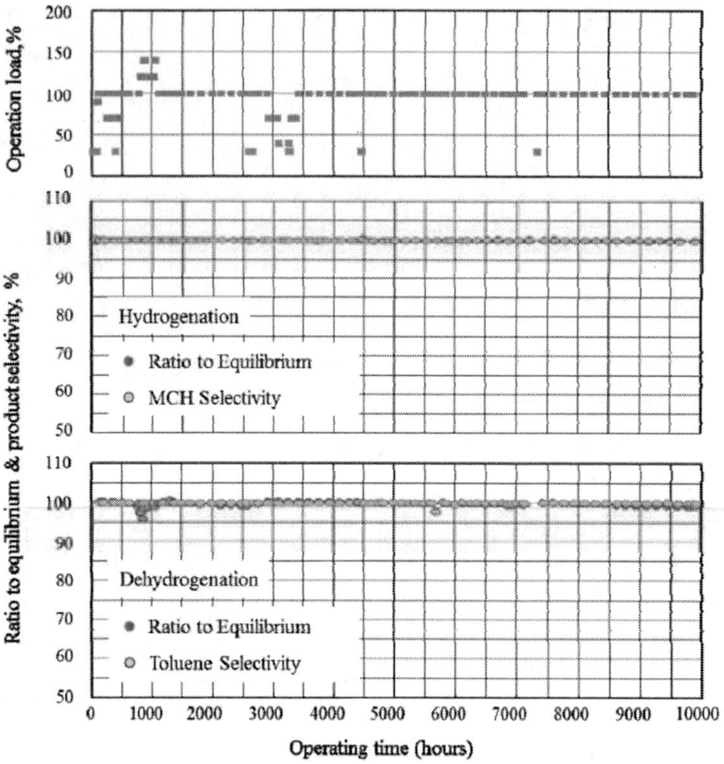

Fig. 6: Performance of hydrogenation & dehydrogenation process

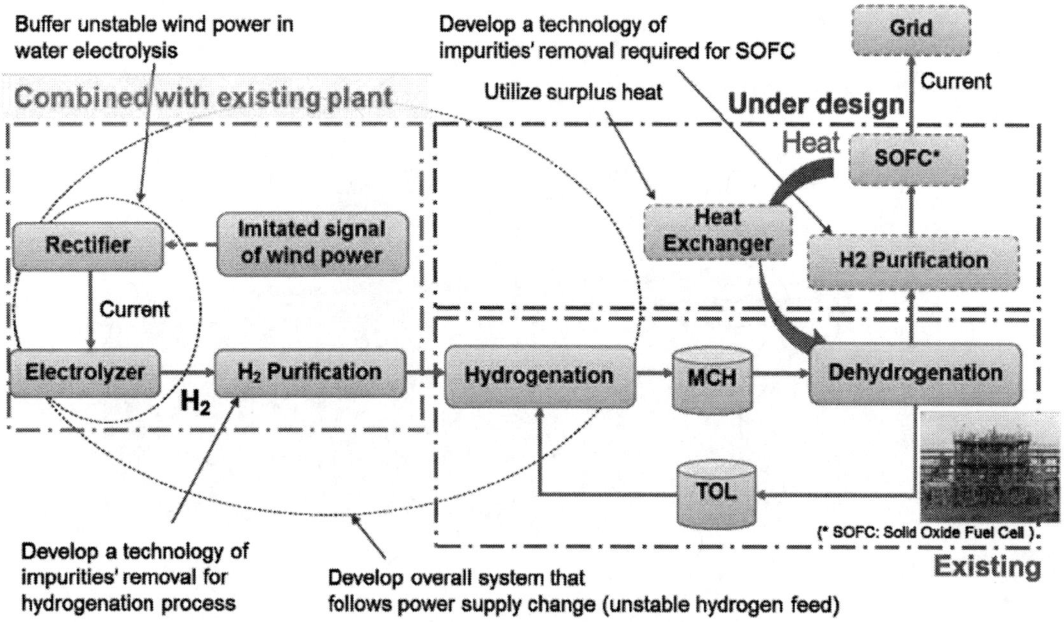

Fig. 7: Demonstration plans for utilizing renewable energies

480396 Exploring the Catalytic Options in Low-Mid Pressure Diesel Hydrotreating

Abstract:

Regularly revised, worldwide diesel fuel specifications continue to press impurity levels lower and product quality higher. Such fuel specification revisions naturally encourage the semi-continuous development of superior hydrotreating catalysts, to satisfy the new requirements and improve the economics of hydrotreacting where applicable. The goal of this talk is to draw rational boundaries where the most active examples of supported NiMo and supported CoMo catalysts ought to be applied, given the inhibiting influences of organic nitrogen species and aromatics species, for example. Included in the scope shall also be application of both types of catalysts simultaneously, where a bed of one of the catalysts is loaded above the other. In consideration of the pressing need for improved hydrotreating catalysts, Honeywell UOP has recently commercialized an entirely new portfolio of supported hydrotreating catalysts under the Unity(TM) trademark. Several of these catalysts shall be featured in this presentation, since they were developed for diesel hydrotreating in the low-to-mid pressure range.

Copyright © American Institute of Chemical Engineers. All rights reserved.

480176 Next-Generation Utilities Optimisation for Refineries and Large-Scale Chemical Production Sites

Abstract:

Refineries and chemical production sites are major consumers of energy in the form of electricity, steam and hydrocarbon feedstocks. Given that tariffs, costs and demands are constantly changing, there is much scope for optimising on-site production, conversion and distribution of energy to minimise cost and emissions, by managing the options available in the most cost effective way while meeting all constraints of the system. However many current 'optimisation' applications do not provide full capabilities for maximising economic value or minimising emissions, particularly in situations where demands from major consumers and the prices of the various components can change on an hourly basis.

This paper describes an advanced optimisation platform for managing and optimising utility operation that not only helps planners rapidly optimise equipment selection and load allocation to improve overall efficiency and reduce emissions and operating costs, but also presents operators with a ranked list of possible actions from which they can choose the best course of action given the situation 'on the ground'.

The approach uses medium-fidelity models of the utilities system and major devices that are coupled with plant operating data via data validation and reconciliation facilities. An advanced optimisation system capable of both continuous and integer decisions determines the economically optimal operating point taking into account equipment and operational constraints, including availability. The resulting mathematical problem is solved within an equation-oriented framework, providing robustness and speed of execution well in advance of most current systems.

Because of the speed of solution, a key feature is the ability to run multiple optimisations and provide operators with a ranked list of potential combinations of changes and their corresponding benefits, within a dashboard tailored for the site. This allows operators clearly to evaluate and discuss which changes are the best to apply when, resulting in advice that is practical and easy to implement and verify. It addresses one of the biggest obstacles to practical realisation of the benefit of optimisation systems, which is gaining operator buy-in to proposed changes.

The paper is illustrated with an examples of implementation on a major European chemical manufacturing site.

Copyright © American Institute of Chemical Engineers. All rights reserved.

478714 Study on Regional Air Quality Impact from Olefin Plant Shutdown Operations

Abstract:

Ground-level ozone is a pervasive air pollutant, which can be potentially aggravated by flaring emissions from olefin plant shutdown operations. Although, flaring is crucial to chemical plant safety, the excessive flaring, for instance the intensive flaring from plant shutdown operations, emits huge amounts of ozone pollution precursors (VOCs and NOx). This will cause tremendous industrial material loss and environmental damage; meanwhile result in potential adverse impact to the regional air quality. Thus, the plant shutdown flare emissions should be minimized at any possibility. In this paper, plant-wide dynamic simulations for flare emissions with regional air-quality modeling are coupled together to quantify the air-quality impact and investigate potential opportunities to mitigate the adverse air-quality impact during plant shutdown operations. It proposed a systematic methodology on air-quality conscious study for shutdown operations of an olefin plant. Through studies of original and optimal cases, it shows that the original plant shutdown operation procedure has a serious and negative effect on the regional air quality; however, the optimal shutdown operation procedure would significantly reduce such impact together with the saving of tremendous emission sources. This study can provide valuable and quantitative supports for decision makings involving multiple stakeholders, including environmental agencies, regional plants, and local communities.

Copyright © American Institute of Chemical Engineers. All rights reserved.

2017 AIChE Spring Meeting & 13th GCPS Global Conference on Process Safety

Up-to-date Engineering Solutions to Estimate and Minimize Thermal Breathing Losses from Storage Tanks

Dr. D. Moncalvo & Dr. M. Davies & Dr. T. Heidermann

Some examples of Tank Vacuum Collapse

Maybe you won't hear about it in the news but it surely costs lots of money!

© 2017 Braunschweiger Flammenfilter GmbH. All rights reserved.

PROTEGO
for safety and environment

Tank Breathing in API 2000 7th Ed. and ISO 28300

Tank breathing in API 2000 7th Ed. and ISO 28300
is distinguished into **Outbreathing** and **Inbreathing**

These standards distinguish between two scenarios:

- Normal in- and out-breathing
  - Liquid pumping
  - Weather changes (sun exposure, rain)
  - Internal failure of heat transfer devices
  - Control valve failures
  - Inert purge and utility failures
  - Vaporization or condensation
  - Steam collapse
  - Other causes
- Emergency out-breathing (fire exposure)

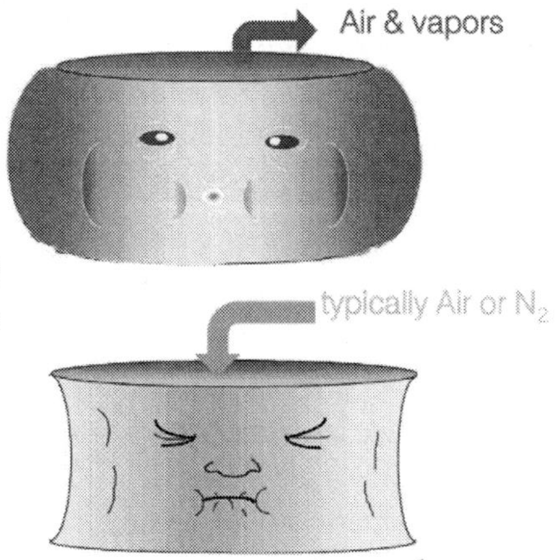

PROTEGO
for safety and environment

© 2017 Braunschweiger Flammenfilter GmbH. All rights reserved.

Tank Thermal Breathing in API 2000 7th Ed. and ISO 28300

$$\dot{V}_{thermal\ out} = Y \cdot V_{tk}^{0.9} \cdot R_i$$

$$\dot{V}_{thermal\ in} = C \cdot V_{tk}^{0.7} \cdot R_i$$

Insulation factor R_i

$$R_i = 1/[1 + h \cdot l_{in}/\lambda_{in}]$$

Not considering condensation or evaporation of product

Latitude		Y Factor	C Factor			
			Vapor pressure			
			Hexane or similar		Higher than hexane or unknown	
			Average storage temperature			
			< 77°F	≥ 77°F	< 77°F	≥ 77°F
0 - 42°	Houston	0.32	4	6.5	6.5	6.5
42° - 58°	Rotterdam	0.25	3	5	5	5
58° - 90°	Norway	0.20	2.5	4	4	4

PROTEGO
for safety and environment

© 2017 Braunschweiger Flammenfilter GmbH. All rights reserved.

Tank Thermal Breathing in API 2000 7th Ed. Annex A (*API 2000 5th Edition*)

Tank capacity	Thermal breathing (*SCFH*)	Max. heat transfer rate
Between 20 000 und 180 000 bbl.	2 SCFH each ft^2 of roof and shall area	Between 50 and 100°F/h
Below 20000 bbl.	1 SCFH each bbl. of tank volume	100°F/h

Annex A method is applicable, if following ***Limitations*** are satisfied:

Tank volume is less than 180 000 bbl.
Max. operating temperature in vapor space is 120°F
Tank is uninsulated
Temperature of stored products and feed is below boiling temperature at max. oper. pressure

Not considering condensation or evaporation of product

© 2017 Braunschweiger Flammenfilter GmbH. All rights reserved.

PROTEGO
for safety and environment

Tank thermal breathing in the scientific literature

Model	Year	Thermal modeling	Evaporation modeling	Condensation modeling	Sensitivity analysis	Measurements
Sigel et al.	1983	X		X		X
Förster et al. (*PTB model*)	1984	X				
Fullarton et al.	1987	X	X	X		X
Holtkötter et al.	1997	X		X		X
Salatino et al.	1999	X				
Brooks (DIERS pres)	2012				X	
Davies and Moncalvo (1)	2016	X		X	X	
Davies and Moncalvo (2)	2016	X	X	X		X

(1)	Journal of Loss Prevention in the Process Industry , volume 43, doi: 10.1016/j.jlp.2016.06.006
(2)	12th Global Congress on Process Safety (GCPS), Houston, April 2016, ISBN: 978-0-8169-1096-0

© 2017 Braunschweiger Flammenfilter GmbH. All rights reserved.

PROTEGO
for safety and environment

Davies and Moncalvo (2016)

- Balance of energy within the tank

$$\left(M_{vap}c_{vap} + M_{liq}c_{liq} + M_{air}c_{air}\right)\frac{dT_B}{dt} = -h_{int}A(T_B - T_E) - \dot{M}_{air}c_{air}(T_B - T_{amb})$$

- Balance of energy at the tank walls

$$M_E c_E \frac{dT_E}{dt} = h_{int}A(T_B - T_E) - h_{rain}A(T_E - T_{rain}) + \rho_B \dot{V}_{cond}\Delta h_{cond}$$

- Balance of mass within the tank

$$-\left(V_{air} + V_{vap}\right)/T_B \cdot dT_B/dt + \dot{V}_{cond} = \dot{V}_{air}$$

- Condensation model

$$\dot{V}_{cond} = \beta \cdot 1/p_B \cdot dp_{sat}/dT \cdot [T_B - T_E]A$$

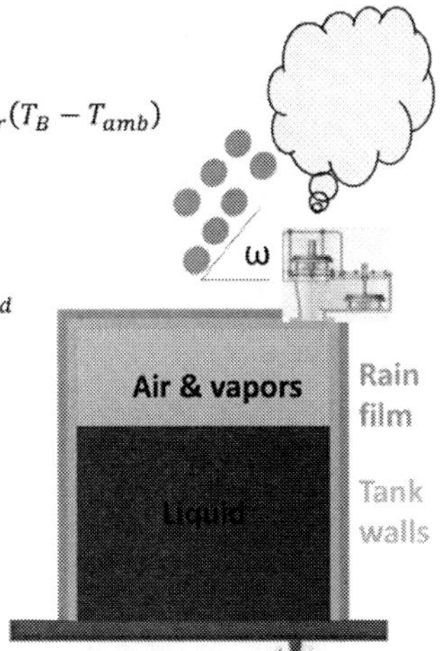

Source	Journal of Loss Prevention in the Process Industry, volume 43, doi: 10.1016/j.jlp.2016.06.006

PROTEGO for safety and environment

© 2017 Braunschweiger Flammenfilter GmbH. All rights reserved.

Some realistic examples of thermal inbreathing

	Tank 1	Tank 2
Tank height	49.2 ft.	
Tank diameter	49.2 ft.	82.0 ft.
Tank volume (V_{tank})	16,650 bbl.	46,250 bbl.
Initial temperature (T_0)	120°F	
Ambient temperature	60°F	
Average storage temperature	68°F	
Rain intensity	46.1 lb/h ft²	
Rain inclination	68°	
Wall material and thickness	Stainless Steel; $^1/_5$ in	
Insulation	Mineral wool	
Stored product	Dry Air Saturated Humid Air Saturated Heptane / Air Saturated Methanol / Air	

PROTEGO for safety and environment

© 2017 Braunschweiger Flammenfilter GmbH. All rights reserved.

Results for ***uninsulated*** tank 1 (D = 49.2 ft.) and tank 2 (D = 82.0 ft.)

Required Inbreathing [SCFH]	API 2000 Main body (max)	API 2000 Main body (min)	API 2000 Annex A	PTB (numerical)	Davies Moncalvo *without* condensation	Davies Moncalvo *with* condensation
Dry air	37,179	23,237	16,659	30,931	31,258	31,258
Humid air				30,730	30,682	48,138
Sat. Heptane / Air				19,576	19,608	31,640
Sat. Ethanol / Air				24,761	24,794	52,825

Required Inbreathing [SCFH]	API 2000 Main body (max)	API 2000 Main body (min)	API 2000 Annex A	PTB (numerical)	Davies Moncalvo *without* condensation	Davies Moncalvo *with* condensation
Dry air	76,013	47,508	35,910	68,807	69,576	69,576
Humid air				67,978	68,149	102,569
Sat. Heptane / Air				41,455	41,680	65,512
Sat. Ethanol / Air				53,485	53,807	108,608

© 2017 Braunschweiger Flammenfilter GmbH. All rights reserved.

PROTEGO
for safety and environment

Worst realistic cases for rain intensity on *Tank 1*

Required Inbreathing [SCFH]		Rain intensity [lb/h ft²]		API 2000 Main body (max)	Davies Moncalvo *without* condensation
Dry air Tank 1	Scenario 1	46.1	API 2000	37,179	31,258
	Scenario 2	62.4	US historical highest (Holt, MO)		34,020
	Scenario 3	85.9	Worldwide hist. highest (Cherrapunji, IND)		36,910

Lesson learned:
API 2000 7th Ed. Main Body is safe for non-condensing products even under extreme weather

© 2017 Braunschweiger Flammenfilter GmbH. All rights reserved.

PROTEGO
for safety and environment

Reducing Tank breathing: Insulation Layer and Tank Wall Thickness on *Tank 1*

Required Inbreathing [SCFH]	Insulation Thickness [in]	API 2000 Main body (max)	Davies Moncalvo without condensation
Dry air Tank 1	uninsulated	37,179	31,258
	1	8,905	1,364
	2	5,058	949
	4	2,714	551

Required Inbreathing [SCFH]	Wall Thickness [in]	API 2000 Main body (max)	Davies Moncalvo without condensation
Dry air Tank 1	0.2	37,179	31,258
	0.25		28,989
	0.3		27,221
	0.4		24,404

Lessons learned:

Increasing tank wall thickness may not suffice to reduce inbreathing requirements

Perform an engineering and economical study to find out the best insulation strategy for your tank

PROTEGO
for safety and environment

© 2017 Braunschweiger Flammenfilter GmbH. All rights reserved.

Tests on the influence of overpressure on tank leakage losses

- Leakage, tested with PROTEGO® portable test rig, may begin at 75% of set pressure as per API 2000 5th Edition Appendix C
- The valves were supposed to be fully open at 1 oz./in² (4.3 mbar) gauge.
- The first test was performed at 100 % overpressure
- The second test was performed at 10% overpressure

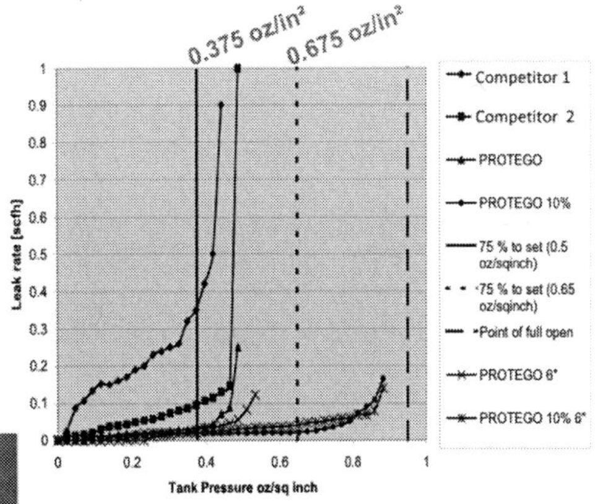

Relieving pressure [oz/in²g]	Overpressure ($\%_{overpr}$)	Set pressure [oz./in² a]	Leakage Pressure [oz./in² g]
1.0	100% (1)	0.5	0.375
1.0	10% (0.1)	0.9	0.675

$$p_{rel\,g} = p_{set\,g}(1 + \%_{overpr})$$

$$p_{leak\,g} = p_{set\,g} \cdot \%_{leak}$$

PROTEGO
for safety and environment

© 2017 Braunschweiger Flammenfilter GmbH. All rights reserved.

Impact of test results in improvement of API 2000 standard

- Two main results from this research test:
  - Avoid choice of valves requiring large overpressures
  - Leakage pressure at 75% set pressure underestimates leakage losses

- Improvement of API 2000 Standard between 5th (1998) and 7th Edition (2014)

Requirements	API 2000 5th Edition	API 2000 7th Edition
Leakage pressure	75 %	90 %
Blowdown	0 %	Dependent on design

- Additional drawbacks to be considered when selecting a PRV with large overpressures:
  - Leakage pressure of PRV may be higher than set pressure of blanketing devices
  - PRV with low set pressures tend to cycle more (additional cost of spare parts)

© 2017 Braunschweiger Flammenfilter GmbH. All rights reserved.

PROTEGO
for safety and environment

Conclusions

API 2000 7th Ed. Main Body is safe for tanks with non-condensing gases even under extreme weather

Do not underestimate condensation in atmospheric tanks and perform an engineering study, especially if your tank contains volatile products, like (m)ethanol or gasoline

Increasing tank wall thickness may not suffice to reduce inbreathing requirements

Perform an engineering and economical study to find out the best insulation strategy for your tank (trade-off between costs of insulation purchase and maintenance against savings on breathing losses)

An effective minimization of product losses suggests the selection of PRV with minimum overpressure: the authors support the 10% overpressure as standard also for PVRV

PVRV with low overpressures usually guarantee also a chatter-free opening and avoid conflict with blanketing valves and ERV.

© 2017 Braunschweiger Flammenfilter GmbH. All rights reserved.

PROTEGO
for safety and environment

Thank you for your kind attention!
Reproduction, in part or in full, subject to prior written approval of Braunschweiger Flammenfilter GmbH.
Braunschweiger
Flammenfilter

480314 Carbon Capture and Reuse in Industrial Scale Continous Biological Processes

Abstract:

Biological processes offer of the lowest energy consuming options for capturing and recycling carbon dioxide into useful sellable or stable products. As tools and technologies continue to advance in transgenics, the possibilities for industrial scale continuous biological processing is becoming a reality. Understanding the state of the art and pathways to address the challenges of scaling-up biological processes for carbon capture and sequestration are critical for advancement of these technologies. This presentation will review the state of the art and evaluate how tools in biochemical engineering (CRISPR) and process engineering will continue to advance biological processes that ultimately rival conventional chemical, petrochemical, and petroleum processes in capability.

Copyright © American Institute of Chemical Engineers. All rights reserved.

479999 Multiple Steady States in Oxidative Steam Reforming of Methanol (OSRM) over CuO-ZnO-Al2O3 Catalyst

Abstract:

Multiple steady states in oxidative steam reforming of methanol (OSRM) over CuO-ZnO-Al2O3 catalyst

Young Shin Jang, Jung Hyun Kim, Jitae Lee, Dong Hyun Kim*

Department of Chemical Engineering, Kyungpook National University

Daegu, 702-701, Korea

1. Introduction

Fuel cell cars are operated on compressed hydrogen. The pressure of the stored hydrogen in the cars is extremely high, around 700 bar, to extend the range between refueling. Nevertheless the range is still relatively short (~ 400 km, Hyundai fuel cell car) compared to ordinary cars using gasoline or diesel. Moreover the fuel cell cars need a new social infrastructure of hydrogen stations. These problems as well as safety issues of handling highly compressed flammable gas act as barriers for widespread use of fuel cell cars. An alternative to storing hydrogen onboard is to produce it onboard from readily available liquid fuel.

Methanol can be regarded as a hydrogen carrier as it can be easily reformed with steam (SRM) to produce hydrogen by

$CH_3OH + H_2O \rightarrow CO_2 + 3H_2$, $\Delta H = 49.5$ kJ/mol (1)

The reaction is endothermic and the SRM reactor has a heat exchange type structure to supply the reaction heat efficiently. On the other hand the reaction heat can be supplied in situ by oxidizing a part of methanol by adding O_2 into the SRM feed. This is called the oxidative steam reforming of methanol (OSRM). As the amount of the heat generation depends on the O_2 content in the feed, an autothermal system (ATR), in which the net enthalpy change is zero, can be made

$CH_3OH + 0.8\ H_2O + 0.10\ O_2 \rightarrow CO_2 + 2.8\ H_2$, $\Delta H = 0$ kJ/mol (2)

In practice, the O_2 content relative to CH_3OH is higher than 0.1 to account for the heat loss of the OSRM reactor. There is still disagreement in the literature on how the O_2 reacts with CH_3OH. Two reactions have been suggested.

$CH_3OH + 0.5\ O_2 \rightarrow CO_2 + 2\ H_2$, $\Delta H = -192.5$ kJ/mol (3)

$CH_3OH + 1.5\ O_2 \rightarrow CO_2 + 2\ H_2O$, $\Delta H = -676$ kJ/mol (4)

Majority of studies on OSRM assumed Eq. (3), partial oxidation of methanol (POM), as the heat generating reaction [1-5], while a few studies have shown that Eq. (4), combustion of methanol (COM), as the oxidation reaction [6,7]. It is one of the objective of this study to experimentally determine which of the two reactions takes place during OSRM.

The exothermic reaction, Eq. (3) or Eq. (4), can give rise to a multiplicity in steady states of OSRM. In the present study, we first report such multiple steady states of OSRM observed experimentally and analyze the states to unravel the heat generation and removal in the OSRM reactor.

2. Experimental

The reactor was made of 10 cm copper tube (3.175 mm OD, 1.55 mm ID). The catalyst in the reactor was a commercial CuO-ZnO-Al_2O_3 (Synetics 33-5) of a size of 0.2 - 0.3 mm (50 mg) diluted with the same volume of glass beads of the same size. The feed rate was 100 ml/min (CH_3OH:H_2O: O_2:He= 15:20:3:62). The reaction temperature was measured at the outer surface of the reactor. The temperature variation along the reactor was negligible because of the high thermal conductivity of the reactor and the short catalyst bed (~ 4 cm). The reaction product was analyzed online with a gas chromatography. The reaction products were mainly H_2 and CO_2 with less than 0.5 % CO.

3. Results and discussions

Fig. 1. shows the methanol conversion with respect to the reaction temperature. With increasing temperature from 160 °C, the conversion followed branch I until 250 °C where the conversion (X) was 0.13 and the products were H_2O and CO_2, showing that COM took place in the branch I. Above 250 °C, the conversion jumped to B at 275 °C where the conversion was 0.77 and with further increase in the temperature, the conversion followed the branch II from B to C. On decreasing the reaction temperature from C, the conversion followed first branch II and then to branch III until D where the reaction temperature was 220 °C and the conversion was 0.19. A further decrease in the temperature from D quenched the reaction to E where T=209 °C and X=0.02.

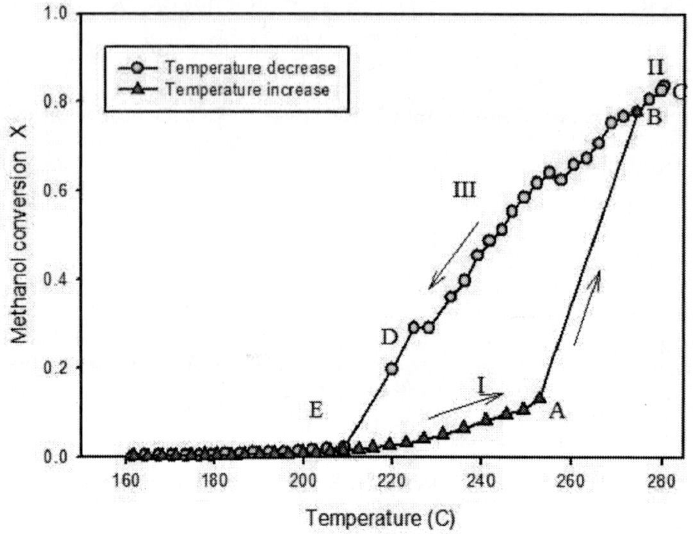

Figure 1. Multiple steady states in oxidative steam reforming of methanol

The O_2 conversion (not shown) was almost 1.0 at A. If O_2 were consumed by POM, Eq. (3), the methanol conversion would have been at least 0.4. The observed X=0.13 at A only corresponds to the combustion of methanol, disproving that POM is an independent single reaction. On branches II and III, the O_2 conversion was always complete and H_2 was produced in proportional to the methanol conversion, showing that SRM followed COM in the catalyst bed.

The multiple steady states were analyzed in terms of heat generation and removal of the reactor [8]. For this purpose, we obtained experimentally the kinetics of COM and the overall heat transfer coefficient between the catalyst bed and its surrounding. The details of the analysis and comparison between the simulation and experiments will be presented at the conference.

4. Conclusions

In oxidative steam reforming of methanol (OSRM), the reaction between oxygen and methanol is experimentally found to be combustion of methanol (COM), Eq. (4), and hydrogen is formed only by the steam reforming of methanol (SRM), Eq. (1). No experimental evidence for the occurrence of the partial oxidation reaction (POM), Eq. (3), has been observed. Multiple steady states of OSRM were experimentally observed and analyzed to be the results of balancing the heat generated by combustion and the heat consumed by steam reforming as well as the heat loss to the reactor surrounding.

References

1. S.Velu , K. Suzuki, M.P. Kapoor, F. Ohashi, T. Osaki, Applied Catal. A: General 213 (2001) 47.
2. J. B. Wang, Chia-Hao Li, Ta-Jen Huang, Catalysis Letters 103(2005).
3. M. Turco G. Bagnasco, U. Costantino, F. Marmottini, T. Montanari, G. Ramis, G. Busca, J. of Catal. 228 (2004) 56.
4. M. Turco, G. Bagnasco, C. Cammarano, P. Senese, U. Costantino, M. Sisani, Applied Catal. B: Env. 77 (2007) 46.
5. S. Patel, K.K. Pant, Applied Catal. A: General 356 (2009) 189.
6. J. Agrell, M. Boutonnet, J. L.G. Fierro, Appl. Catal. A: General, 253 (2003) 213.
7. M. Lyubovsky, S. Roychoudhury, Appl. Catal. B: Environmental, 54 (2004) 203.
8. H. Fogler, Elements of Chemcal Reaction Engineering, 4th ed. Prentice Hall (2006) 533.

Copyright © American Institute of Chemical Engineers. All rights reserved.

480965 Technology Advancement in the Field of Ethylene Oxide and Ethylene Glycol

Abstract:

SABIC is the world largest MEG producer in petrochemical sector and most of its MEG is being produced through the hydrolysis of EO. Ethylene oxide (EO) is a simplest form of epoxides and due to its molecular structure, it acts as a versatile intermediate for a variety of chemicals such as; monoethylene glycol (MEG), di- and triethylene glycols, ethoxylates, ethanol amines, and glycol ethers. The growing demand of MEG owing to its downstream utilization in polyester industry has prompted the research of an efficient, cost effective, and environmentally benign EO production process.

Currently, most of the World's MEG is being produced through conventional route; however, there is an increasing trend to exploit new routes, and feedstocks such as coal, bio-based, and Shale gas. The production capacity of MEG is expected to increase to 36 million tons by 2024. The heavy dependence of MEG producers on ethylene and its scarcity in some parts of the world promoted the need of alternate feedstock.The technological as well as commercial aspects will be discussed along with the future growth of EO/EG business.

Copyright © American Institute of Chemical Engineers. All rights reserved.

481850 Environmental Impact of Mining

Abstract:

Mining can have adverse effects on surrounding surface and groundwater if protective measures are not taken. The result can be unnaturally high concentrations of some chemicals, such as arsenic, sulfuric acid, and mercury over a significant area of surface or subsurface.[3] Runoff of mere soil or rock debris -although non-toxic- also devastates the surrounding vegetation. The dumping of the runoff in surface waters or in forests is the worst option here. Submarine tailings disposal is regarded as a better option (if the soil is pumped to a great depth).[4] Mere land storage and refilling of the mine after it has been depleted is even better, if no forests need to be cleared for the storage of the debris. There is potential for massive contamination of the area surrounding mines due to the various chemicals used in the mining process as well as the potentially damaging compounds and metals removed from the ground with the ore. Large amounts of water produced from mine drainage, mine cooling, aqueous extraction and other mining processes increases the potential for these chemicals to contaminate ground and surface water. In well-regulated mines, hydrologists and geologists take careful measurements of water and soil to exclude any type of water contamination that could be caused by the mine's operations. The reducing or eliminating of environmental degradation is enforced in modern American mining by federal and state law, by restricting operators to meet standards for protecting surface and ground water from contamination. This is best done through the use of non-toxic extraction processes as bioleaching. If the project site becomes nonetheless polluted, mitigation techniques such as acid mine drainage(AMD) need to be performed.

The five principal technologies used to monitor and control water flow at mine sites are diversion systems, containment ponds, groundwater pumping systems, subsurface drainage systems, and subsurface barriers. In the case of AMD, contaminated water is generally pumped to a treatment facility that neutralizes the contaminants.[5]

A 2006 review of environmental impact statements found that "water quality predictions made after considering the effects of mitigations largely underestimated actual impacts to groundwater, seeps, and surface water".[6]

Copyright © American Institute of Chemical Engineers. All rights reserved.

THE POTENTIAL USE OF SOLAR PHOTOCATALYTIC OXIDATION FOR REMOVAL OF EMERGING PHARMACEUTICALS FROM EFFLUENT OF WASTEWATER TREATMENT PLANT: A PILOT PLANT STUDY

Fares Almomani, Rahul Bhosale, Anand Kumar, Majeda Khreisheh, Ujjal Gosh
Department of Chemical Engineering, Qatar University, P.O box 2713, Doha, Qatar

Abstract

The treatment of effluent wastewater containing pharmaceutical by solar advanced oxidation processes were studied in a pilot plant. The removal efficiency of four groups of pharmaceuticals (antibiotics, estrogens, acidic, and neutral), commonly detected as emerging contaminants, were tested by different solar-driven photochemical processes including solar -photolysis, solar-photocatalytic oxidation using Fe(III) or TiO_2 and solar based photo-Fenton reaction. Results showed that Solar-photolysis is not effective for the complete removal of the selected contaminants, on other hand solar-photocatalytic oxidation using TiO_2, solar based photo-Fenton reaction showed significant removal efficiencies and high mineralization.

Solar photo-catalytic systems produced a final solution with lower toxicity if compared with original solution.

Keywords: photons, emerging contaminates, solar excitation, hydroxyl radicals

Introduction

Incident presence of what is called emerging contaminates(ECs)such as pharmaceuticals, pesticide in water and effluents of wastewater treatment plants have raised a lot of concern regarding the performance of conventional treatment processes in removing and treating these contaminants (Ternes, 2001, Kolpin et al., 2002). It was reported that the main reason for the presence of these ECs is related to the continual release of these contaminates into this eco-system from different point sources (Huber et al., 2004). ECs are considered non-biodegradable, and consequently it is expected that their concentration in the environment will continue to increase unless they are effectively treated and eliminated. Reasonable concentration of these contaminates have been detected in secondary treatment effluents of wastewater treatment plants (WWTPs) (Kolpin et al., 2002). Contamination of groundwater with ECs was reported from different sources(Ternes et al., 2002). Thus, there is a big need for an effective and cheap post-treatment to remove ECs from wastewater before discharge to the environment.

ECs can be effectively removed from water solution using different treatment processes such as activated carbon, nano-filtration and reverse osmosis(Rosenblatt et al., 1967) (Reddersen et al., 2002, Webb et al., 2003). However, the cost associated with these treatment alternatives limits their applicability.

Advanced oxidation processes (AOPs), on other hand, can be employed as an effective and promising technology for the removal of any traces of ECs from water and wastewater. AOPs which based on the production of highly oxidative radicals are considered as destructive processes. These processes have the capacity to mineralize different contaminates from wastewater in a short period of time. With such technologies, problematic compounds can be partially or totally oxidized into smaller by-products that can be easily removed in conventional biological processes. Applying this strategy would potentially reduce the size of both oxidation and biological reactors, and may reduce both capital and operating costs. This synergy has the potential to open up a new niche in the water treatment field for treating contaminants which are conventionally reserved for other treatment technologies (e.g. activated carbon). The high cost associated with the use of AOPs (reagent and UV-light) limit their applicability. However, most recently, solar advanced oxidation processes were effectively used in the treatment of water and wastewater containing non-biodegradable and/or toxic organic pollutant (Almomani et al, 2010, Al

Momani and Jarrah, 2009). Using solar irradiation as a source of photon can reduce the cost of AOPs and make them a suitable treatment options.

The present work aims to investigate the suitability of using solar induced AOTs for removal of ECs from water and wastewater. The study will be carried out in a compound parabolic collector (CPC) prototype pilot plant solar reactor. The CPC technology allows all the received solar radiation to be utilized for oxidation. Thus, oxidation performance increase. The outcome of this project will allow to develop a new technology for the treatment of water and wastewater containing ECs by use of solar energy which is abundant in Qatar and the Arab gulf region.

Materials and Methods

Experimental Set Up

Experiments set up used to carry out pharmaceutical solar oxidation is shown in (Fig. 1). The solar pilot plant is equipped with (1) a wastewater containing pharmaceutical. The tank was immersed in water bath to control the working temperature, (2) compound parabolic collector (CPC) solar reactor . the CPC allows for almost all the solar irradiation received at the reactor aperture area (not only direct, but also diffuse) to be collected being available for oxidation. The solar consists of six quartz tubes with a length of 50 cm, an inner diameter of 3 cm and outer diameter of 3.1 cm. This pilot plant has the capacity to treat up 5 L of wastewater, the solution was circulated through the reactor with a flow rate of 20 L. h^{-1} and the sun's movement was tracked manually.

Analytical methods

The pharmaceutical concentrations were measured with a Hewlett-Packard 1050 series HPLC equipped with an Ultra Aqueous C_{18} column from Restek (3.2100 mm, 5 mm) and a variable wavelength detector. The eluents consisted of 10 mM phosphoric acid (A) and pure acetonitrile (B). The eluent flow rate was 0.5mL.min^{-1} and the detection limit was 10 μg.L^{-1}. The 95% confidence intervals for a single measurement were typically 3–5%.

The dissolved organic carbon (DOC) was measured using a Shimadzu TOC analyzer. The analyzer was calibrated using potassium phthalate standard solution and following the instrument standard method. Each sample was analyzed three times and measurements were averaged. Measurements reproducibility showed a precision in the range of ±0.1 mg.C.L^{-1}.

Water samples

Wastewater (WW) was collected from the effluent of secondary treatment train from a Doha north WWTP. Water samples were filtered through a 0.45 μm Millipore filter, and stored at 4 °C prior to use. Table 1 shows the characteristics of Wastewater treatment plant effluent.

Table 2: Water quality parameters for surface and wastewaters.

Parameter	Wastewater treatment plant effluent Average, (std. dev.)[1],
pH	7.9 (0.4)
DOC ($mg.L^{-1}$)	13(1.2)
Alkalinity ($mgCaCO_3.L^{-1}$)	145 (6.0)
Nitrate ($mg\text{-}N.L^{-1}$)	31 (1.4)

Nitrite (mg-N.L^{-1})	15 (0.4)
Sulfate (mg.L^{-1})	50 (2.0)
Chloride (mg.L^{-1})	15 (1.2)

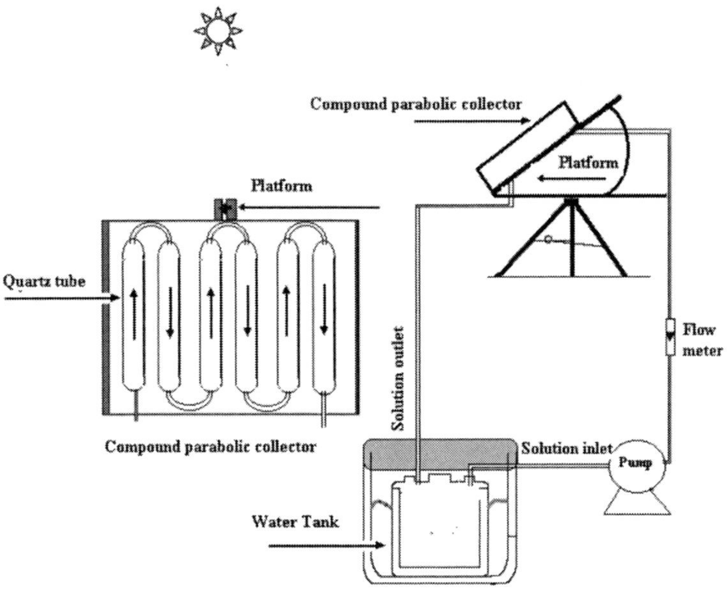

Figure 1: Solar Experimental Setup

Experimental Procedure

Experiments were conducted Doha, Qatar (25.2854° N, 51.5310° E) in a semi-batch mode, under sunlight and at ambient temperature. At the beginning of the reactor was loaded with 5 L of the aqueous solution of pharmaceuticals and the required amount of catalyst was added afterwards. Then, the solution was recirculated through the CPC by a centrifugal pump at a flow rate of 20 $L.hr^{-1}$ so that the flux pattern inside the tubes was laminar (the Reynolds number was 2500). For the first 10 min, the CPC tubes were kept covered to avoid sunlight illumination. This time was enough to reach the adsorption equilibria of pharmaceuticals onto TiO_2 (if used). After that, the CPC was uncovered and the reaction was was started. Also, if needed, hydrogen peroxide was added at this point. Samples were withdrawn at intervals through a sampling port to determine the concentration of pharmaceuticals as function of reaction time, the remaining total organic carbon (TOC), and the concentrations of hydrogen peroxide, ozone and Fe(II). The ozone concentration in the outlet gas stream was also continuously monitored with in-line ozone analyzers.

The initial concentration of pharmaceuticals used in experiment are presented in table 1. The initial concentrations were selected to improve the chemical analysis accuracy and facilitate mineralization kinetic study. The initial Fe(III) and TiO_2 concentrations were 3 and 100 mg L^{-1}, respectively.

Toxicity analysis

Toxicity bioassays were performed to evaluate the toxicity of water containing pharmaceuticals before and after ozonation. Tests were carried out using Microtox® bioassays, which involved the use of the marine bacterium *Photobacterium phosphoreum* as an indication of sample toxicity. The standard Microtox protocol was followed, and data analysis was achieved using the Microtox™ Omni Software (SDI, 2002). Sample toxicity was evaluated at 5 and 15 min from the time of mixing of the sample with marine bacterium at various dilutions. The toxicity was determined by measuring the EC_{50} of the sample. The EC_{50} is the concentration of the test sample at which there is 50% reduction in bacterial light production. This value was related to sample toxicity using the following equation:

$$TU = \frac{100}{EC_{50}} \quad [1]$$

where TU is the toxicity unit and EC_{50} is the concentration of the test sample at 50% reduction in bacterial light production.

Results and Discussion

Solar -Photolysis of **pharmaceuticals**

Figure 2 A and B show the evaluation of pharmaceutical as a function of solar radiation time during solar photolysis process. Direct solar photolysis was not effective in removing or degradation of pharmaceuticals from wastewater. This is an expected fact as the solar radiation with wavelength < 290 nm does not reach Earth's surface as it is mostly retained by atmosphere [32]. In addition, all the target pharmaceuticals do not absorb radiation above 300 nm significantly. The removal efficiencies ranged from 9.1- 12.1%, 13-15.6%, 13.9 -16 and 5.4 and 8.2 for antibiotics, estrogen, acidic and neutral pharmaceutical, respectively.

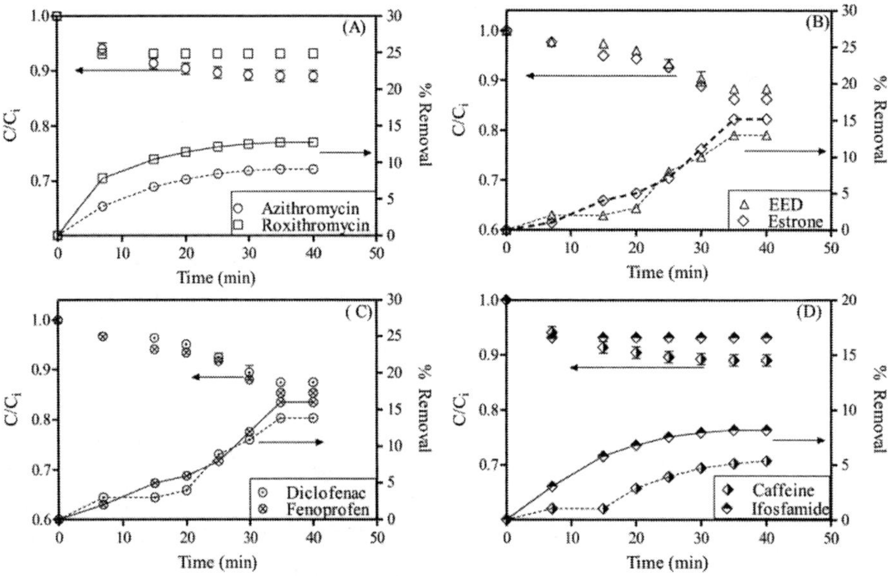

Figure 2: Evolution of the normalized concentration $[C]/[C]_i$ and % removal of antibiotics, Estrogens, Acidic and natural pharmaceuticals during solar photolysis

Solar-photocatalytic oxidation of pharmaceuticals

Iron (Fe(III) and titanium dioxide were used as catalyst for this part of experiments. Figure 2 shows the results for experiments carried out with iron as catalyst in the AOPs including Fe(III)/solar-hv and Fe(III)/solar-hv/H_2O_2. The reactions involved in these oxidation processes are summarized in equations (1 and 2).

$$Fe^{3+} + H_2O \rightleftarrows Fe(OH)^{2+} + H^+ \quad \text{(Reaction 1)}$$

$$Fe(OH)^{2+} \xrightarrow{h\nu} Fe^{2+} + \cdot OH$$ (Reaction 1)

During the photocatalytic oxidation with Fe(III)/hv Pharmaceuticals oxidation occurs by the formation of aqua complexes, mainly $Fe(OH)_2^+$ (see reaction (1)), which can undergoes photolysis under UV radiation ($\lambda > 300$ nm) to generate hydroxyl radicals as shown reaction 2. As it can be seen in Figure 3, the combination Fe(III)/hv led to reasonable degradation of the pharmaceuticals. The removal efficiencies for antibiotics, estrogen, acidic and neutral pharmaceutical ranged from 74 – 77.8%, 69-71.9%, 70.8-80.9% and 57.6-66.8, % respectively. The results suggest that solar-photocatalytic oxidation can be used to oxidize these pharmaceuticals. However, the reaction time should be increase of there is a need to add reagent to assist in oxidation.

Adding H_2O_2 to Fe(III)/hv change the system a well-known oxidation system called photo-Fenton. In this case, H_2O_2 accelerates the production of hydroxyl radicals as in reactions (3) and (4) and change Fe (III) to Fe(II) (reaction 5) which, in turn, can reduce Fe(III) to Fe(II) as in reaction (6). The net results is more hydroxyl radicals and more pharmaceuticals oxidation.

$$Fe^{2+} + H_2O_2 \rightarrow Fe^{3+} + OH^- + \cdot OH$$ (Reaction 3)

$$H_2O_2 \xrightarrow{h\nu} 2\cdot OH$$ (Reaction 4)

$$H_2O_2 + \cdot OH \rightarrow HO_2^{\cdot} + H_2O$$ (Reaction 5)

$$Fe^{3+} + HO_2^{\cdot} \rightarrow O_2 + Fe^{2+} + H^+$$ (Reaction 6)

Solar photo catalytic oxidation by TiO_2 is well-known and it has been widely used to oxidize non-biodegradable organic matter in water systems.

The change in solution toxicity after different oxidation process were evaluated. For experiments carried out with photolysis, it was observed that oxidized water samples toxicity does not change a suggesting that the low oxidation reaction did not change water toxicity. For water samples with higher oxidation such as photo-Fenton and TiO_2/hv system lower acute toxicity than the initial wastewater was observed. As an example, water containing 3.75 mg/L of Roxithromycin showed a TU value of 17, oxidation photo-Fenton and TiO_2/hv system decreased the TU to 1 and 1.2, Respectively. Similar results were observed for other pharmaceutical. The reduce in toxicity is related to the reaction between radicals and the studied pharmaceuticals at tertiary amines bonds leading to the cleavage of the N–C bonds. This reaction can lead to either the destruction of the chemical structure of pharmaceuticals or the loss of the methyl group from the amino group. Li et al., 2001 have confirmed that the loss of the methyl group of roxithromycin eliminates its apparent activity. Besides, the oxidation reaction can cause cleavage of the phenolic structure in pharmaceuticals and reduce its estrogenicity significantly. On the basis of these considerations, solar photocatalytic oxidation of pharmaceuticals results in significant reduction in pharmacological effects of these pharmaceuticals

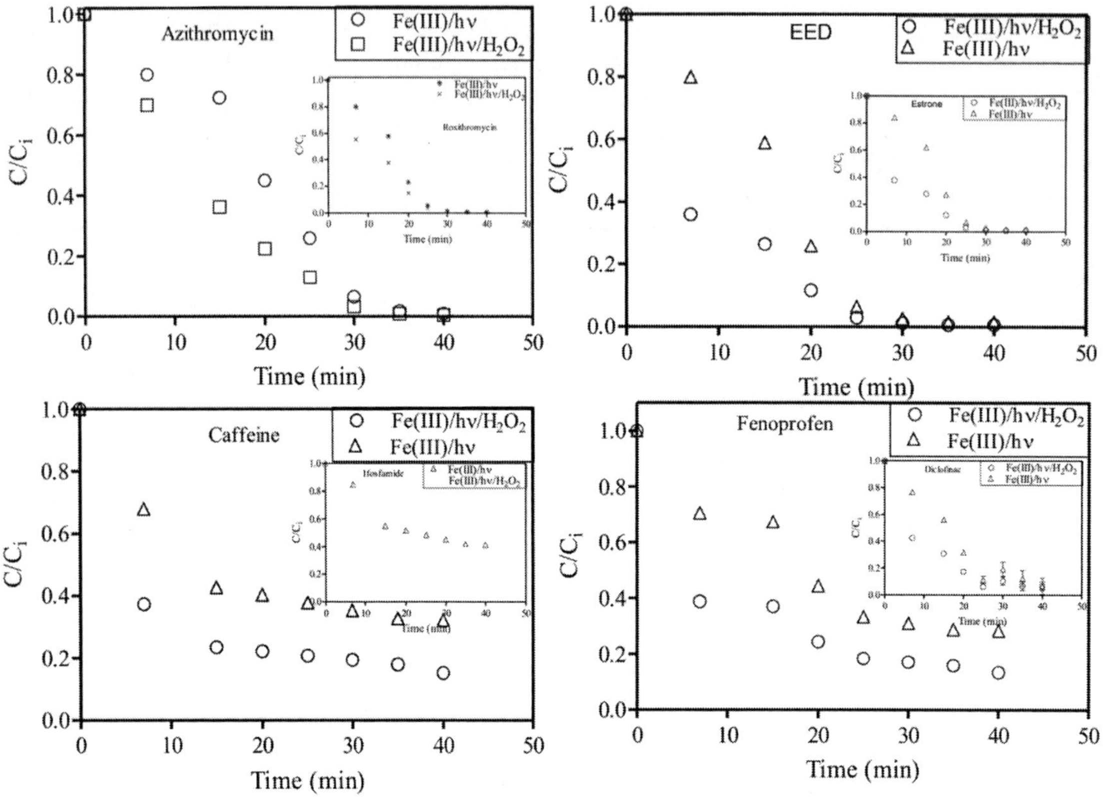

Figures 3 : Evolution of the normalized concentration [C]/[C]$_i$ of antibiotics, Estrogens, Acidic and natural pharmaceuticals during solar-photocatalytic oxidation.

Conclusion:

Oxidation of four groups of pharmaceuticals (antibiotics, estrogens, acidic and neutral) with solar advanced oxiadation was investigated in effluent of WWTP. Solar photocatalytic reaction can be used successfully to eliminates nob-biodegradablke pharamacuticals from WWTP effluets. The degradation rate of pharmaceuticals was strongly accelerated by adding catalys or hydrogen peroxide to the system. 98 % of pharmaceutical degradation was achieved by solar photo-Fenton or photo-catakytic oxidation with TiO_2in 40 min reaction time.

Toxicity tests using Microtox^{TM} bioassay were completed to evaluate the effect of the oxidation reactor on the toxicity of water solution before and after the treatment. Solar oxidation reduced the a cute toxicity of all the pharmaceuticals in the wastewater suggesting that this technology can be coupled with wastewater treanment process to complete the removal of the toxic and non-biodegradable materials

Acknowledgements

This publication was made possible by the NPRP grant (NPRP No.: 6 - 1436 - 2 - 581) from the Qatar National Research Fund (a member of Qatar Foundation). The statements made herein are solely the responsibility of authors. The authors gratefully acknowledge the support provided by the Qatar University to do this work..

REFERENCES

DODD, M. C. & HUANG, C.-H. (2004) Transformation of the Antibacterial Agent Sulfamethoxazole in Reactions with Chlorine: Kinetics, Mechanisms, and Pathways. *Environmental Science & Technology,* 38, 5607-5615.

GAMAL EL-DIN, M., SMITH, D. W., AL MOMANI, F. & WANG, W. (2006) Oxidation of resin and fatty acids by ozone: Kinetics and toxicity study. *Water Research,* 40, 392-400.

HUBER, M. M., KORHONEN, S., TERNES, T. A. & VON GUNTEN, U. (2005) Oxidation of pharmaceuticals during water treatment with chlorine dioxide. *Water Research,* 39, 3607-3617.

HUBER, M. M., TERNES, T. A. & VON GUNTEN, U. (2004) Removal of Estrogenic Activity and Formation of Oxidation Products during Ozonation of 17alpha -Ethinylestradiol. *Environmental Science & Technology,* 38, 5177-5186.

KOLPIN, D. W., FURLONG, E. T., MEYER, M. T., THURMAN, E. M., ZAUGG, S. D., BARBER, L. B. & BUXTON, H. T. (2002) Pharmaceuticals, Hormones, and Other Organic Wastewater Contaminants in U.S. Streams, 1999-2000: A National Reconnaissance. *Environmental Science & Technology,* 36, 1202-1211.

LOVATO, M. A. E., MARTÃ-N, C. A. & CASSANO, A. E. (2009) A reaction kinetic model for ozone decomposition in aqueous media valid for neutral and acidic pH. *Chemical Engineering Journal,* 146, 486-497.

PINKSTON, K. E. & SEDLAK, D. L. (2004) Transformation of Aromatic Ether- and Amine-Containing Pharmaceuticals during Chlorine Disinfection. *Environmental Science & Technology,* 38, 4019-4025.

REDDERSEN, K., HEBERER, T. & DANNBIER, U. (2002) Identification and significance of phenazone drugs and their metabolites in ground- and drinking water. *Chemosphere,* 49, 539-544.

ROSENBLATT, D. H., HULL, L. A., DE LUCA, D. C., DAVIS, G. T., WEGLEIN, R. C. & WILLIAMS, H. K. R. (1967) Oxidations of Amines. II. Substituent Effects in Chlorine Dioxide Oxidations. *Journal of the American Chemical Society,* 89, 1158-1163.

TERNES, T. (2001) Pharmaceuticals and Metabolites as Contaminants of the Aquatic Environment. *Pharmaceuticals and Care Products in the Environment.* American Chemical Society.

TERNES, T. A., MEISENHEIMER, M., MCDOWELL, D., SACHER, F., BRAUCH, H.-J. R., HAIST-GULDE, B., PREUSS, G., WILME, U. & ZULEI-SEIBERT, N. (2002) Removal of Pharmaceuticals during Drinking Water Treatment. *Environmental Science & Technology,* 36, 3855-3863.

WEBB, S., TERNES, T., GIBERT, M. & OLEJNICZAK, K. (2003) Indirect human exposure to pharmaceuticals via drinking water. *Toxicology Letters,* 142, 157-167.

ZHANG, P. & LIU, J. (2004) Photocatalytic degradation of trace hexane in the gas phase with and without ozone addition: kinetic study. *Journal of Photochemistry and Photobiology A: Chemistry,* 167, 87-94.

481851 Oil Pollution, Environmental Degradation and Restoration of Nigeria's Ogoniland

Abstract:

Human activities especially those of oil exploration and exploitation raise a number of issues such as depletion of biodiversity, coastal and riverbank erosion, flooding, oil spillage, gas flaring, noise pollution, sewage and wastewater pollution, land degradation, soil fertility loss and deforestation, which are all major environmental issues.

Ogoniland is characterized by typically deltaic features; uneven terrain, numerous creeks, shallow brackish water bodies and a variety of vegetation types including swamp forest. The land has a groundwater either as localized aquifers or in hydraulically interconnected aquifers. Ogoniland is not an Island. This has two implications. The first is that pollution from Ogoniland has the potential to reach and cross its boundaries as well as entering Ogoniland from extreme sources.

Petroleum hydrocarbons are naturally occurring hydrocarbons substances and depending on the length of the carbon, can occur in gas, liquid and solid form. However, hydrocarbons refer to chemical substances formed exclusively from carbon and hydrogen by the decay of organic substances trapped within sedimentary rocks.

Hydrocarbons substantially alter the physical and chemical properties of the soil once it comes in contact with it. The contact could be from the natural seepage of hydrocarbons to accidental seepage of crude oil on the ground. In the case of water, hydrocarbons can cause both physical and chemical effects by preventing oxygen transfer in the water column, thus affecting aquatic life-support systems. As regards vegetation, where spillage is not immediately attended to, oil spills often lead to fires, causing total or partial destruction of vegetation. Dissolved or emulsified oil in the water column can contaminate the plankton, algae, fish, eggs and the vertebrae larvae. Also, physical contacts with oil destroys the insulation of fur and feathers, and more often, birds ingest oil which have lethal or sub-lethal impacts through, for example, liver and kidney damage.

Residents in these areas, face the risk of dangerous exposure to hydrocarbons when they breathe, bathe, eat fish, drink water or accidentally eat or touch soil or sediment that is contaminated with oil. In addition to the chemical pollution by hydrocarbons, there are other environmental concerns linked with oil industry operations. These range from clearance of land for oilfield facilities, hydrological changes due to construction of roads and pipelines and contamination from chemicals other than hydrocarbons.

There must be maintenance of oilfield facilities, decommissioning of oilfield facilities and prevention of illegal activities, oil spill response and remediation of contaminated sites. Environmental restoration includes the establishment of an Integrated Contaminated Soil Management Centre (ICSMCS), setting up of mini-treatment centers and rehabilitation of mangroves.

Part of the actions is the amendment of the official gazette establishing the Hydrocarbon Pollution Restoration Project (HYPREP), to reflect a new governance framework comprising a Governing Council, a Board of Trustees and Project Management.

There must be an end to the widespread pipeline sabotage, crude oil theft and illegal refining that are the main cause of environmental damage in Ogoniland and the wider Niger Delta.

There should be transfer of oversight of the Environmental Guidelines and Standards for Petroleum Industry in Nigeria (EGSPIN), to ensure constant surveillance. Also, there should be the provision for social and health impact assessment as an integral part of the overall environmental impact assessment (EIA) process for all new oil and gas facilities and upgrades to existing facilities in line with international best practice. This means that the Environmental Impact Assessment (EIA) Act should be amended to accommodate these changes. It also proposes that there should be a review of the provisions of the National Oil Spill Detection and Response Agency NOSDRA (Establishment) Act, 2006 against NOSDRA's current operational responsibilities. Oversight of clean-up should be given to a separate governmental department.

Copyright © American Institute of Chemical Engineers. All rights reserved.

480271 Selecting Right Flare System to Meet Regulatory Requirments

Abstract:

Flare is considered as last line of defense in any process plant. This is one equipment which must be available at all time when the plant is in operation including during emergency cases. The purpose of the flare system is to reliably and safely dispose any and all hydrocarbon releases from the plant during startup, normal operation, maintenance and emergencies. The Federal Regulation (40 CFR 60.8) requires that emissions, except for emergency cases, shall be smokeless. 'Smokeless' flare operation is defined as generating emissions within 20% opacity levels, also known as Ringelmann#1 (R1) on Ringelmann smoke Chart. Typically, this smokeless flow tends to fall within 10 to 15% of the emergency flaring load for an LNG (Liquefied Natural Gas) plant. The apparent darkness or opacity of a stack plume depends upon:

- Concentration and size of the particulate matter in the effluent
- Depth of the smoke column being viewed
- Natural lighting conditions such as the direction of the sun relative to the observer, and Color of the particles

There are several limitations to select the right flare system. In the case study presented in this paper, four major criteria have been considered for evaluation. (a) total plot space availability / usage, (b) Flaring noise during normal operation, start-up & maintenance cases (c) Equipment cost and (d) Ease of maintenance. Smokeless characteristic depends on many factors including turbulence (characterized by exit velocity and mixing), molecular weight, heating value and chemical structure of waste gases and time available for combustion.

In the case study presented in paper for an LNG plant, the following type of flares are evaluated to optimize the flare selection, design, and cost.

- Elevated pipe flare including air / steam assisted flaring.
- Multi-point Ground flare (MPGF).
- Totally Enclosed Ground Flare or Enclosed Ground Flare (TEGF or EGF).

The project team needs to get a very good estimate of the amount of plot required for the flare system even in the Front End Engineering phase of the project. A detailed flare study is required to determine the type and cost of the flaring system.

This paper discusses various flare system designs available to achieve smokeless flaring during normal operation, start-up, and maintenance. Selection of right flare or right combinations of flare types are very important and must be evaluated carefully for overall plant reliability and cost effectiveness.

Copyright © American Institute of Chemical Engineers. All rights reserved.

480334 Optimization of Shutdown Procedures of Olefin Plants

Abstract:

Shutdown and startup procedures are important operations for maintenance of olefin plants. Both procedures involve complicated facility operations, take long time and may produce huge amount of flare emissions. Optimize shutdown procedures could shorten duration time, reduce flare emissions and decrease cost by saving intermediate materials.

In this presentation, a new shutdown methodology for olefin plants is proposed. The key issue is to recover remaining liquid and vapor materials in facilities during the shutdown procedure. To achieve that, CGC (Charge Gas Compressors) must be kept running during recovering, and recovered hydrocarbon is sent to other plants as feed. Dynamic simulations of the whole procedure are used to determine possibility and to provide necessary operation data.

Moreover, the shutdown procedure can be considered as a scheduling problem. Thus, an MIP (Mixed-Integer Programming) model is developed to get an optimized shutdown schedule for facilities. By optimizing shutdown arrangements for facilities, the overall shutdown time could be reduced by 25%. Combining with the new shutdown methodology, over 90% of hydrocarbon could be saved, and flare emissions could also be reduced by 90%.

Copyright © American Institute of Chemical Engineers. All rights reserved.

AUTHOR INDEX